HIRING HOUSEHOLD HELP AND CONTRACTORS: YOUR OBLIGATIONS UNDER THE LAW

by

Margaret C. Jasper

Oceana's Legal Almanac Series:
Law for the Layperson

2005
Oceana Publications, Inc.
Dobbs Ferry, New York

Library of Congress Control Number: 2005925301

ISBN 0-379-11396-1

Oceana's Legal Almanac Series: Law for the Layperson
ISSN 1075-7376

Manufactured in the United States of America on acid-free paper.

To My Husband Chris

Your love and support
are my motivation and inspiration

-and-

In memory of my son, Jimmy

Table of Contents

CHAPTER 2:
PERFORMING A BACKGROUND CHECK

CHAPTER 3:
EMPLOYMENT VERIFICATION REQUIREMENT

CHAPTER 4:
INSURING YOUR HOUSEHOLD HELP

CHAPTER 5:
EMPLOYMENT TAXES

APPENDICES

ABOUT THE AUTHOR

MARGARET C. JASPER is an attorney engaged in the general practice of law in South Salem, New York, concentrating in the areas of personal injury and entertainment law. Ms. Jasper holds a Juris Doctor degree from Pace University School of Law, White Plains, New York, is a member of the New York and Connecticut bars, and is certified to practice before the United States District Courts for the Southern and Eastern Districts of New York, the United States Court of Appeals for the Second Circuit, and the United States Supreme Court.

Ms. Jasper has been appointed to the panel of arbitrators of the American Arbitration Association and the law guardian panel for the Family Court of the State of New York, is a member of the Association of Trial Lawyers of America, and is a New York State licensed real estate broker and member of the Westchester County Board of Realtors, operating as Jasper Real Estate, in South Salem, New York. Margaret Jasper maintains a website at http://www.JasperLawOffice.com.

Ms. Jasper is the author and general editor of the following legal almanacs: AIDS Law; The Americans with Disabilities Act; Animal Rights Law; The Law of Attachment and Garnishment; Bankruptcy Law for the Individual Debtor; Individual Bankruptcy and Restructuring; Banks and their Customers; Buying and Selling Your Home; The Law of Buying and Selling; The Law of Capital Punishment; The Law of Child Custody; Your Rights in a Class Action Suit; Commercial Law; Consumer Rights Law; The Law of Contracts; Copyright Law; Credit Cards and the Law; The Law of Debt Collection; Dictionary of Selected Legal Terms; The Law of Dispute Resolution; Drunk Driving Law; DWI, DUI and the Law; Education Law; Elder Law; Employee Rights in the Workplace; Employment Discrimination Under Title VII; Environmental Law; Estate Planning; Everyday Legal Forms; Executors and Personal Representatives: Rights and Responsibilities; Harassment in the Workplace; Health Care and Your Rights. Home Mortgage Law Primer; Hospital Liability Law; Identity Theft and How To Protect Yourself; Insurance Law; The Law of Immigration; International Adoption; Juvenile Justice and Children's Law; Labor Law; Landlord-Tenant Law; The Law of Libel and Slander; Living Together: Practical Legal Issues; Marriage and Divorce; The Law of Medical Malpractice; Motor Vehicle Law; The Law of No-Fault Insurance; Nursing Home Negligence; The Law of Obscenity and Pornography; Patent Law; The Law of Personal Injury; The Law of Premises Liability; Prescription Drugs; Privacy and the Internet: Your Rights and Expectations Under the Law; Probate Law; The Law of Product Liability; Real Estate Law for the Homeowner and Broker; Religion and the Law; Retirement Planning; The Right to Die; Rights of Single Parents; Law for the Small Business Owner; Small Claims Court; Social Security Law; Special Education Law; The Law of Speech and the First Amendment; Teenagers and Substance Abuse; Trademark Law; Victim's Rights Law; The Law of Violence Against Women; Welfare: Your Rights and the Law; What if it Happened to You: Violent Crimes and Victims' Rights; What if the Product Doesn't Work: Warranties & Guarantees; Workers' Compensation Law; and Your Child's Legal Rights: An Overview.

INTRODUCTION

Whether you need a nanny to watch your children, or a nurse to care for your elderly parent, hiring employees to work in your home makes you an employer. As an employer, you have a number of legal responsibilities that must be fulfilled concerning your household employees.

Many people are not aware of the legal responsibilities that go along with hiring employees to work in their homes. There are numerous and complex immigration laws, labor laws, and tax laws that must be understood and followed. Even though the work takes place in your home, it is nonetheless employment, and the people you hire are your employees under the law.

This almanac examines the law as it pertains to hiring and managing your household employees. The household employer's obligations concerning employment eligibility verification, payroll taxes, wages, benefits, and insurance are also set forth in this almanac.

This almanac also discusses the differences between hiring employees and hiring independent contractors, and the criteria one must consider in making this important distinction. The precautions one must take when hiring a contractor to perform services, including selecting the right contractor for the job, and negotiating a contract for their services, is also explored.

The Appendix provides sample forms, and other pertinent information and data. The Glossary contains definitions of many of the terms used throughout the almanac.

CHAPTER 1: THE HOUSEHOLD EMPLOYER

SHOULD I HIRE HOUSEHOLD HELP?

In today's society, where both parents often work, hiring household help to handle time-consuming tasks has become a necessity instead of the luxury it was once considered. According to recent studies, housework can take up 35 hours of precious time each week. It's a full-time job in itself. And, for parents who work, this leaves little time to spend with the family, and to just relax.

Under this type of schedule, some important aspect of your life will suffer as a result, such as your job performance or your family life. So, if you find yourself in this situation, the answer to the question is yes, you should hire some household help to ease your burden, whether it's a cook to prepare the evening meal, or somebody to clean your windows.

In deciding what type of help would be most cost-effective and beneficial to your situation, you should take into account certain factors. For example, what tasks are the most time-consuming? If you find yourself spending hours per day washing the family's clothes, you may want to hire someone to do laundry and light housework.

If there are chores that you really hate to do, like walking the dog or mowing the lawn, you should consider hiring someone to perform those services, and leave the more tolerable tasks for yourself. Time is money, and hiring someone to free up your time for more important things is well worth the cost.

Generally, household work is work done in and around your home by the following people:

1. Babysitters
2. Cleaning people

3. Drivers
4. Housekeepers
5. Nannies
6. Health aides
7. Private nurses
8. Maids
9. Caretakers
10. Yard workers
11. Other similar types of domestic help

Child care workers who perform their services for you in their own home are generally not considered employees.

HOW DO I FIND HOUSEHOLD HELP?

Hiring a person to work in your home is a very personal decision. After all, your home is your sanctuary—the place where you and your family can relax and feel secure. Your home is the place where you are entitled to utmost privacy. Your most precious possessions are located in your home. Therefore, allowing strangers into your home is a serious decision which requires careful consideration and good judgment.

One of the best ways to find household help is to ask friends and relatives for referrals. People you trust make very good references, and it is unlikely that they will send an unsuitable employee your way. You can also place an advertisement in your local newspaper which sets forth all of the requirements of the position. Often, household employees will place their own advertisement in the newspaper indicating their availability.

SCREENING PROSPECTIVE EMPLOYEES

It is important to diligently screen and personally interview all prospective employees. As part of your hiring process, you should draft a detailed application, and verify all of the information contained in the application. Following are suggested topics you should discuss during the personal interview:

1. Prior employment, including names and contact information for former employers, dates of employment, and responsibilities in the position.

2. If the applicant is being hired to care for children, or for an elderly or incapacitated family member, you should inquire as to their feel-

ings about caring for children, disabled persons, the elderly, etc. You want to try and determine how they would handle temper tantrums, stubbornness, anger, etc.

3. If your children, or elderly parent, is available, you should introduce them and try to determine whether they are a good "match." Watch their interactions, assess whether the applicant is comfortable dealing with them. Obtain feedback from your family members and carefully consider their feelings. After all, they are the ones who will be spending the most time with the prospective employee.

4. If the applicant is going to be responsible for shopping, transporting children, etc., you must find out if they have a reliable car, their driving history, etc.

5. You must discuss the applicant's availability and make sure he or she can work the days and hours required.

6. You should present the applicant with a detailed job description, as discussed below, and make sure they are able to perform the services contained in the job description. For example, if they will be required to prepare meals, is cooking a skill that they possess?

7. Find out if the applicant has any personal habits that would conflict with your lifestyle, e.g., smoking.

8. Ask open-ended questions that give the applicant the opportunity to demonstrate their personality, qualities, skills, etc.

9. You should ask for both work-related and personal references.

10. Consider a trial period before you hire the applicant for a full-time permanent position. Check all of the references provided by the applicant.

Be sure that you have a chance to watch the interactions between the in-home worker and the family member for whom he or she will be providing care. You may want to do this at the end of the interview with individuals you feel are good candidates, or you may want to invite the top two candidates back to meet with your family member. If your family member is able, he or she should be included in the interview process and in making the final decision.

In addition to carefully screening the prospective employee, you should undertake a thorough background check before hiring. If you are unable to do the background check yourself, you should hire a reputable investigation company to do the work for you.

Background checks are discussed more fully in Chapter 2.

HIRING THROUGH AN EMPLOYMENT AGENCY

If you are unable to find suitable referrals, you can go through an employment agency. Before you contact an agency, however, you should investigate the agency's references, and check with the Better Business Bureau to see whether any complaints have been made against the agency.

There are many agencies that place household employees. Using an outside agency is particularly important if you need a professional to work in your home, such as a nurse or physical therapist. In that case, you should select an agency that provides employees who are experienced in their field.

You should also obtain complete information on how the agency finds the employees it places. You must make sure that the agency carefully screens prospective employees. Are applicants personally interviewed by the agency? Does the agency carefully check the employees' references? Is the employee required to fill out an application and, if so, is all of the information contained on the application verified by the agency? You want to make sure that the agency doesn't take the information contained on an application at face value.

Also find out if the agency conducts background checks of prospective employees, and to what extent they investigate the employees. For example, do they check on whether a the applicant has a criminal record or poor driving record? Ask whether the agency conducts drug testing of prospective employees.

You must ask the agency whether the person you are hiring is your employee or the employee of the agency. Some agencies function as the employer. You pay the agency, and the agency pays the household employee. You do not have an employer-employee relationship with that person. Other agencies charge a one-time fee for placing a household employee, and are no longer responsible for that employee once placed. Thereafter, you function as the employer and pay the employee directly.

If the household worker is the employee of the agency, ask the agency for a copy of its certificate of insurance. The certificate of insurance provides documentation that the agency provides workers' compensation insurance for its employees. Also inquire as to whether the agency provides health and disability coverage for the employee.

PREPARING AN EMPLOYMENT CONTRACT

It is important to treat your household employees as employees. This means that you should prepare a job description which details the type of position for which you are hiring the employee, the tasks you want the employee to perform, and the employee's responsibilities. You should then prepare an employment contract formalizing the conditions of employment.

The employment contract should contain, at a minimum, the following provisions:

1. Identification information for the household employee, including their name, address, telephone number, social security number, etc.

2. Wages, including when and how payment will be made, e.g., weekly, bi-weekly, etc.

3. Benefits, such as holidays and vacations.

4. The employee's schedule, including days and hours the employee is expected to work.

5. The duties the employee is being hired to perform, as set forth in the job description.

6. Actions or behaviors that will result in termination, e.g., theft, lateness, etc.

7. Performance evaluations.

8. Termination provisions.

Both the employer and employee should sign and date the employment contract, and a copy of the contract and any additional documents, such as the job description, should be given to the employee for his or her records.

A sample household employer/employee employment agreement is set forth at Appendix 1.

EMPLOYMENT TERMINATION

Unless your employment agreement with your household employee states that the employee is guaranteed employment for a certain term, e.g., one year, an employer can usually discharge or lay off an employee, with or without cause, and with or without notice. In turn, an employee is entitled to quit their job without repercussion, absent a provision in the employment agreement. This is known as "employment at will."

Although an employer basically has the right to hire and fire employees at will, the law does provide some restrictions. As set forth below, an employer cannot discriminate in making any adverse employment decision, e.g., on the basis of age, race, gender, religion, national origin, or disability.

RISK MANAGEMENT

A developing area of risk management law affecting household employers is known as Employee Practices Liability (EPL). It is not uncommon for a disgruntled employee to bring a frivolous action against a well-to-do employer. Many of these lawsuits are brought based on discrimination in hiring, or wrongful termination. Employers who have a large household staff are more likely to be the target of such a lawsuit simply because there are more employees being hired and fired.

Whether or not a lawsuit is without basis, they are nonetheless time-consuming and costly. You can minimize your risk of being the victim of a groundless lawsuit by taking certain precautions.

First, it is important to keep proper records. Be sure to document each employee's hours and wages to guard against claims of wage and overtime violations. You should also develop an employee handbook containing all of the rules and regulations of employment. As set forth above, you should also prepare a job description and employment contract, which both you and the employee should sign and date.

In addition, you should conduct periodic employee evaluations so that if you have to fire an employee who performs poorly, you have documentation to back up your reasons for the termination. Have each employee sign their evaluation to demonstrate that they were on notice concerning the information contained in the evaluation.

In addition, in order to avoid an employment termination lawsuit, many employers will negotiate certain benefits, e.g. severance pay, in return for an employee's waiving their right to sue.

As discussed in Chapter 4, you can also purchase EPL insurance coverage to shield yourself from personal financial liability in case you are the subject of such a claim.

HIRING MINOR EMPLOYEES

It is not uncommon to hire a teenage boy to mow your lawn, or a teenage girl to care for your children. However, you must be aware of the laws pertaining to employment of individuals under the age of 18 so that you do not inadvertently violate the law. The Fair Labor Standards Act (FLSA) sets 14 as the minimum age for most non-agricultural work performed by children, with some exceptions. Children aged 18 years or older may perform any job, and are no longer subject to the child labor laws.

Many states have enacted child labor laws, some of which may have a minimum age for employment which is higher than the FLSA. Where both the FLSA and state child labor laws apply, the higher minimum standard must be obeyed.

A table of minimum age requirements for employment, by state, is set forth at Appendix 2.

Type of Work Permitted by Age

The type of work a child is allowed to perform depends on the age of the child, as set forth below.

All Children - Any Age

Under the FLSA, all children, even those under the age of 14, may deliver newspapers; perform in radio, television, movie, or theatrical productions; work in businesses owned by their parents, except in mining, manufacturing or hazardous jobs; and perform babysitting or perform minor chores around a private home.

Children Aged 14 and 15

Under the FLSA, in addition to the above, children aged 14 and 15 can work in an office, grocery store, retail store, restaurant, movie theater, baseball park, amusement park, or gas station, and can perform the following kinds of tasks:

1. Bagging and carrying out customer's orders;

2. Cashiering, selling, modeling, art work, advertising, window trimming, or comparative shopping;

3. Cleaning fruits and vegetables;

4. Clean-up work and grounds maintenance. The young worker may use vacuums and floor waxers, but he or she cannot use power-driven mowers, cutters, and trimmers;

5. Delivery work by foot, bicycle, or public transportation;

6. Kitchen and other work in preparing and serving food and drinks, but not cooking or baking;

7. Office and clerical work;

8. Pricing and tagging goods, assembling orders, packing, or shelving;

9. Pumping gas, cleaning and polishing cars and trucks (but the young worker cannot repair cars, use garage lifting rack, or work in pits);

10. Wrapping, weighing, pricing, stocking any goods as long as the young worker does not work where meat is being prepared and does not work in freezers or meat coolers.

Children Aged 16 and 17

Under the FLSA, children aged 16 and above can work in any job or occupation that has not been declared hazardous by the Secretary of Labor, as set forth below.

Limitations on Work Hours

In addition to the type of work a child is allowed to perform, the number of hours a child is permitted to work also depends on the age of the child, as set forth below.

Children Aged 14 and 15

Under the FLSA, hours worked by 14 and 15 year-olds are limited to:

1. Non-school hours;
2. Three (3) hours in a school day;
3. Eighteen (18) hours in a school week;
4. Eight (8) hours on a non-school day;
5. Forty (40) hours on a non-school week; and
6. The hours between 7 a.m. and 7 p.m., except from June 1 through Labor Day, when evening hours may be extended to 9 p.m.

Children Aged 16 and Above

The FLSA does not limit the number of hours or times of day for workers 16 years and older.

Employment Eligibility Certification

The FLSA does not require children to obtain employment certificates—also referred to as work permits or working papers—to get a job. Some states, however, do require the minor to have an employment certificate prior to employment.

A table of employment certificate issuance practices under state law is set forth at Appendix 3.

An employer may request a minor employee to provide an age certificate in order to assure the employer he or she is in compliance with the child labor laws. Minors can usually obtain an age certificate from school authorities, which the employer can then keep on file as conclusive proof of the worker's age.

EQUAL EMPLOYMENT OPPORTUNITY LAWS

Except for the Equal Pay Act (EPA), most household employers are not subject to the Equal Employment Opportunity laws (EEO Laws) because these laws generally pertain only to private employers that employ 15 or more individuals. However, because some larger estates may employ more than 15 individuals, the laws are briefly discussed below.

Covered Employers

All employees, including part-time and temporary workers, are counted for purposes of determining whether an employer has a sufficient number of employees to be covered under the EEO laws. An employee is someone with whom the employer has an employment relationship. Independent contractors are not counted as employees. Determining whether an individual is an employee or independent contractor is discussed in Chapter 6 of this almanac.

Title VII of the Civil Rights Act of 1964 (Title VII)

Title VII of the Civil Rights Act of 1964 (Title VII) covers all private employers that employ 15 or more individuals.

The Americans with Disabilities Act (ADA)

The Americans with Disabilities Act (ADA) cover all private employers that employ 15 or more individuals.

The Age Discrimination in Employment Act (ADEA)

The Age Discrimination in Employment Act (ADEA) covers all private employers with 20 or more employees.

Discriminatory Practices

Under Title VII, the ADA, and the ADEA, it is illegal to discriminate in any aspect of employment on the basis of race, color, religion, sex, national origin, disability, or age (40 years of age or older). It is also illegal to harass employees on the same bases.

An employee who wins his or her case can recover compensatory and punitive damages in cases of intentional discrimination, and attorney's fees.

Covered employers are required to post notices to all employees advising them of their rights under the laws EEOC enforces and their right to be free from retaliation. Such notices must be accessible, as needed, to persons with visual or other disabilities that affect reading.

Discrimination Under The Immigration Reform and Control Act (IRCA)

As set forth in Chapter 3 of this almanac, The Immigration Reform and Control Act (IRCA) requires employers to assure that employees hired are legally authorized to work in the U.S. However, an employer who requests employment verification only for individuals of a particular national origin, or individuals who appear to be or sound foreign, may violate both Title VII and IRCA. Verification must be obtained from all applicants and employees. Employers who impose citizenship requirements or give preferences to U.S. citizens in hiring or employment opportunities also may violate IRCA.

The Equal Pay Act (EPA)

The Equal Pay Act (EPA) covers all employers who are covered by the Fair Labor Standards Act (FLSA). Virtually all employers are subject to the provisions of this Act. The EPA prohibits discrimination on the basis of sex in the payment of wages or benefits, where men and women perform work of similar skill, effort, and responsibility for the same employer under similar working conditions.

THE EMPLOYEE'S RIGHT TO FAIR PAY

An employee's right to receive fair pay and other related matters is also governed by the Fair Labor Standards Act (FLSA). The FLSA sets forth the applicable federal minimum wage and overtime requirements. The FLSA generally applies to household workers, such as nannies and housekeepers. Independent contractors are not covered under the FLSA as they are not deemed to be employees.

Covered employers are required to abide by the FLSA unless a state or other federal law requires a higher standard. The reader is advised to check the law of his or her jurisdiction to ascertain whether an applicable state law provides greater employee protection.

Under the FLSA, an employee is entitled to a minimum hourly wage for his or her services. The reader is advised to also check his or her state's minimum wage laws, because federal minimum wage laws defer to the state's minimum wage law, provided the state sets a higher rate than the federal government.

In addition, an employee is entitled to overtime pay. Overtime pay must be at least one and one-half times the employee's regular rate of pay for all hours worked over 40 in a workweek.

Fines of up to $1,000 per violation may be assessed against employers who willfully or repeatedly violate the minimum wage or overtime pay provisions. In addition, the Department of Labor may recover back wages for employees who have been underpaid in violation of the law, and those employers found to be in violation may face civil or criminal action against them.

A directory of State Departments of Labor is set forth at Appendix 4.

CHAPTER 2:
PERFORMING A BACKGROUND CHECK

IN GENERAL

Before you hire an applicant, you must also undertake a thorough background check of the prospective employee. However, before you start investigating the applicant's background, find out about the laws in your state regarding background checks. Your state may require disclosure to the applicant that a background check will be performed. To protect yourself, it is advisable to have the applicant sign a release permitting you to go forward with the background check.

CRIMINAL RECORD

As part of the background check, you must find out whether the applicant has a criminal record. Obviously, you do not want anyone who has been convicted of theft to work in your home and have access to your belongings, nor do you want anyone who has been convicted of child endangerment, neglect or abuse working around your children.

In addition, it is a sad fact that there are many sexual predators in our society. When hiring a household employee, you must check to see whether the applicant is on any sex offender registry, particularly if his or her responsibilities include caring for young children.

The bottom line is that you should conduct as extensive investigation as your level of comfort requires. After all, your privacy and safety are at risk.

CIVIL LAWSUITS

You should also find out whether there are any relevant civil lawsuits involving the applicant, such as lawsuits by or against a former employer. If so, get the details of the litigation.

DRIVING RECORD

Finding out whether the applicant has a clean driving record is also important, particularly if the employee will be driving your children back and forth to school and activities. Hiring an applicant who has drunk or reckless driving convictions, or numerous accidents on record, would obviously place the safety of your loved ones at risk.

The National Driver Register (NDR), is a computerized database of information about drivers who have had their licenses revoked or suspended, or who have been convicted of serious traffic violations such as driving while impaired by alcohol or drugs. State motor vehicle agencies provide the NDR with the names of individuals who have lost their driving privilege, or who have been convicted of a serious traffic violation. When a person applies for a driver's license the state checks to see if the name is in the NDR files. If a person has been reported to the NDR as a problem driver, the license may be denied.

If driving is a responsibility of your prospective employee, you may wish to request an NDR file check. In order to obtain this information, the employee may go to the local motor vehicle agency and ask for an NDR file check. For this purpose, an NDR File Request form must be completed by the employee and submitted to the state in which the employee is licensed. Any information sent to the employer from the NDR should also be given to the employee.

If the employee does have an NDR file, the NDR will provide the employer with the name of the state, and the address and telephone number in order to request a copy of the driver record to verify that it is the same individual. Any information on the NDR file that was reported by the states during the past 3-years will be disclosed.

A copy of the Employee NDR File Request Form is set forth at Appendix 5.

For general information on the National Driver Register, contact:

Highway Safety Programs
National Driver Register
400 7th Street SW, Washington, DC 20590-0001
Tel: (202) 366-4800
Website: http://www.nhtsa.dot.gov/people/perform/driver/

A directory of state National Driver Register contacts is set forth at Appendix 6.

CREDIT CHECK

You may also perform a credit check on the prospective employee. This is especially important if the prospective employee will be handling the family's money. In addition, a bad credit history may indicate a lack of responsibility on the part of the applicant. Contact information for the three major credit reporting agencies is as follows:

EQUIFAX
P.O. Box 740241
Atlanta, GA 30374
Phone: (800) 685-1111
Website: www.equifax.com

EXPERIAN
National Consumer Assistance Center
P.O. Box 2002
Allen, TX 75013
Phone: (888) EXPERIAN
Website: www.experian.com/consumer/

TRANS UNION LLC
Consumer Disclosure Center
P.O. Box 1000
Chester, PA 19022
Phone: (800) 888-4213
Website: www.transunion.com

LIE DETECTOR TESTING

In making a decision to hire, many employers would like the right to conduct a lie detector test of every prospective employee. In addition, employers would also like to be able to force existing employees to submit to a lie detector test if the need arises, e.g. where there is suspicion of internal theft. However, under the Employee Polygraph Protection Act, most private employers are prohibited from using lie detector tests in either situation, with limited exceptions.

In those limited situations where lie detector tests are legally permitted, the employee has certain rights relating to the length and conduct of the test. In addition, the employer is required to give the employee advance written notice. An employee may refuse to take the test or may request to discontinue taking the test during its administration. The results of an employee's lie detector test are private and are not permitted to be disclosed to any unauthorized person.

Violations of the Act may result in civil penalties up to $10,000. Legal action may be brought by a job applicant, an employee, or by the Secretary of Labor. In addition, where a specific state law provides for a more restrictive ban on lie detector tests, the federal government defers to that law.

DRUG AND ALCOHOL TESTING

It is advisable to request a prospective employee to submit to drug and alcohol testing as a condition of employment. In general, state and federal laws have been passed to allow employers to test their employee for on-the-job drug or alcohol abuse. Presently, 32 states have enacted employment drug or alcohol testing laws. Because state laws vary, the reader is advised to check the law of his or her own jurisdiction for specific information.

HIRING AN INVESTIGATION COMPANY

There are companies that will conduct a complete background search of a prospective employee for a fee. Due to the availability of the internet, and the amount of information that can be obtained in this manner, a thorough background check can usually be conducted economically and expeditiously.

CHAPTER 3: EMPLOYMENT VERIFICATION REQUIREMENT

IN GENERAL

The U.S. Citizenship and Immigration Services (USCIS), formerly the INS, requires all employers, including household employers, to make sure their employees are legally entitled to work in the United States. Aliens must have authorization from the USCIS to work in the United States. If they work without such authorization, they are subject to deportation for violating their status.

Prior to the enactment of the Immigration Control and Reform Act of 1986 (IRCA), employers were generally not penalized for employing unauthorized aliens. Since IRCA was enacted, employers are now subject to sanctions for failure to abide by the regulations concerning employment of aliens, as discussed below.

VERIFICATION REQUIREMENT

Under IRCA, an employer is prohibited from knowingly hiring an alien not authorized to work in the United States. This includes a ban on: (a) hiring an alien knowing he or she is unauthorized to work; (b) continuing to employ an alien knowing that he or she is, or has become, unauthorized to work; and (c) hiring any person, whether a citizen or alien, without verifying their documentation, as required.

An employer must verify the documentation of any job applicant who is subsequently hired, whether or not the employer was certain the applicant was a U.S. citizen. Even though citizens and nationals of the United States are automatically eligible for employment, they are also required to present proof of employment eligibility and identity to prospective employers.

Citizens of the United States include persons born in the 50 states, Puerto Rico, Guam, the U.S. Virgin Islands, and the Northern Mariana Islands. Nationals of the United States include persons born in American Samoa, including Swains Island.

EMPLOYMENT ELIGIBILITY VERIFICATION FORM (FORM I-9)

Once the employer decides to hire a particular applicant, the employee must complete the employee section of an Employment Eligibility Verification Form (Form I-9) which attests to his or her authorization to work in the United States.

The employer must complete the employer section of the form, by examining documents provided by the employee as evidence of his or her identity and employment eligibility, and attesting to the document examination on the form. The type of documentation necessary to meet verification is prescribed by law. Employees must present original documents. The only exception is a certified copy of a birth certificate.

The employer must examine the documents and, if they reasonably appear on their face to be genuine and relate to the person presenting them, the employer must accept them, otherwise the employer is in danger of committing an unfair immigration-related employment practice. However, if a document does not reasonably appear on its face to be genuine and relate to the person presenting it, the employer must not accept it.

An employer who knowingly uses or accepts fraudulent documentation to satisfy the verification requirement is subject to civil liability, including the imposition of fines.

If an employee is hired with a temporary work authorization, reverification must take place on or before the expiration date of the temporary work authorization.

A sample Employment Eligibility Verification Form (Form I-9) is set forth at Appendix 7.

The employer must maintain the I-9 forms in their files until (a) three years after the date of hire; or (b) one year after termination, whichever comes later. The employer does not have to submit the document to the USCIS or any other government agency. However, the form must be available for review upon notice by an authorized government official.

A completed Form I-9 is not required for applicants who are not subsequently hired. For purposes of the Form I-9 requirement, a person is "hired" when he or she begins to perform labor or services in return for wages or other compensation, also referred to as "remuneration."

Remuneration is anything of value given in exchange for labor or services rendered by an employee, including food and lodging. Thus, in the case of a household employee, a nanny who is hired to care for your children in exchange for free room and board would be considered an employee, and is thus required to complete Form I-9.

EXCEPTIONS TO THE VERIFICATION REQUIREMENT

An employer can terminate an employee who fails to produce the required documentation—or a receipt for replacement documents in the case of lost, stolen or destroyed documents—within three business days from the date employment began. If an employee has presented a receipt for replacement documents, the employee must produce the actual documents within 90 days from the date employment began. However, there are exceptions to the verification requirement including, but not limited to, the following:

1. The Grandfather Clause - An exception to verification exists under IRCA's "grandfather clause," which provides that unauthorized aliens who were hired prior to IRCA's enactment on November 6, 1986 are exempt from the law. Thus, the employment status of these workers does not have to be verified.

2. Rehiring - An exception to verification exists if the employer rehires the same employee within three years of the original hiring date and completion of Form I-9, provided the employee's form indicated that the employee was authorized to work.

3. Referral from State Job Service - An exception to verification exists if the employee is referred by a state job service that: (a) undertook the verification procedure, and (b) provided copies of their documentation to the employer.

USCIS INVESTIGATION

USCIS inspectors are empowered to inspect an employer's recordkeeping files without issuing a subpoena or obtaining a warrant. In fact, noncompliance with a request to review the employer's I-9 forms file may subject the employer to sanctions.

In addition to reviewing the I-9 forms, the USCIS inspectors may request consent to speak with employees, and to review additional employee files. If consent is denied, the USCIS inspector may issue an administrative subpoena to compel production of witnesses and/or records.

The employer is entitled to a three-day notice of an impending investigation. However, if the USCIS has probable cause to believe a violation exists, notice is not required.

SANCTIONS

If an employer fails to verify the documentation of any job applicant who is subsequently hired, the employer is subject to fines even if no unauthorized aliens were hired. An employer who violates the prohibition on hiring unauthorized aliens is subject to severe civil and criminal penalties. The fines range from $250 to $10,000 for each unauthorized alien, depending on how many times the employer has broken the law.

If Form I-9 is properly completed, and the employer subsequently finds out that the employee was not actually authorized to work, the employer cannot be charged with a verification violation. In this case, an employer has a good faith defense against the imposition of employer sanctions for knowingly hiring an unauthorized alien unless the government can prove the employer had actual knowledge of the unauthorized status of the employee. However, once the employer finds out that the employee was not authorized to work, the employer cannot knowingly continue to employ the individual.

Further, an employer who fails to maintain the required records is subject to civil fines ranging from $100 to $1,000 for each individual for whom records are not maintained. More severe penalties may be assessed if it is determined that the employer has engaged in a "pattern or practice" of violating the regulations.

The size of the fine for recordkeeping violations is determined by several factors, including but not limited to: (a) the employer's good faith; (b) the employer's history of recordkeeping violations; (c) the seriousness of the violation; (d) the size of the business; and (e) whether the employee for whom the records were not maintained is, in fact, an unauthorized alien.

EMPLOYMENT AUTHORIZATION DOCUMENTS

An alien who is not a lawful or conditional permanent resident may need to apply for an Employment Authorization Document (EAD) to prove to an employer that they are authorized to work in the United States. A lawful permanent resident or a conditional permanent resident does not need an EAD. Their Alien Registration Card proves that the employee may work in the United States. In addition, an alien who is authorized to work for a specific employer does not need an EAD.

Their passport and Arrival-Departure Record proves that they may work in the United States.

The USCIS issues Employment Authorization Documents for, among others, asylees and asylum seekers; refugees; students seeking particular types of employment; applicants to adjust to permanent residence status; people in or applying for temporary protected status; fiances of American citizens; and dependents of foreign government officials.

The applicant should not let their EAD expire, and should apply for a renewal EAD six months before their original EAD expires. If the EAD becomes lost, stolen or mutilated, or if the information on the original EAD is incorrect, an applicant may apply for a replacement EAD with the USCIS. If the USCIS does not approve or deny an EAD application within 90 days of filing, the applicant may request an interim EAD. If the alien is an asylee, the time within which the USCIS must approve or deny the EAD application is shortened to 30 days.

If the applicant's request for an EAD is denied, he or she will receive a letter that explains why the application was denied. The applicant is not allowed to appeal a negative decision to a higher authority. However, they may submit a motion to reopen or a motion to reconsider the decision with the office that denied the application. By filing these motions, the applicant asks the office to reexamine or reconsider their decision.

A motion to reopen must state the new facts that are to be provided in the reopened proceeding and must be accompanied by affidavits or other documentary evidence. A motion to reconsider must establish that the decision was based on an incorrect application of law or USCIS policy, and further establish that the decision was incorrect based on the evidence in the file at the time the decision was made.

The Immigration and Nationality Act (INA) Section 274A governs the administration of Employment Authorization Documents. The Code of Federal Regulations (CFR) discusses the employment authorization responsibilities of both employers and employees.

CHAPTER 4: INSURING YOUR HOUSEHOLD HELP

IN GENERAL

When you hire household help, you must make sure you get the appropriate insurance coverage to shield yourself from personal financial liability.

AGENCY EMPLOYEES

As set forth in Chapter 1, if you contract with an outside agency for the services of a worker, such as a nurse, cook, or other household employee, you must determine whether that person is your employee or the employee of the agency. If he or she is an employee of the agency, you must ask them for a copy of their certificate of insurance in order to document that the agency provides workers compensation, health and disability insurance for its employees. If so, you are pretty much covered in case the worker is injured on your property.

LIABILITY INSURANCE

Before you hire household help, you should check your homeowner's insurance policy to find out whether they are covered under your homeowner or renter policy in case they are injured, and whether the coverage is adequate. For example, if your babysitter falls down the stairs and breaks her leg, who will pay for her medical expenses? Ask your insurance agent whether your policy includes a no-fault provision which would pay the medical bills directly.

AUTOMOBILE INSURANCE POLICY

If your employee's responsibilities include driving your car, e.g., to run errands, transport your children, etc., you must make sure that you notify your insurance company that there is an additional driver.

DISABILITY INSURANCE

In many states, household employees pay state disability insurance as a part of the taxes withheld from their paycheck. Therefore, if they must take time away from work due to illness or maternity leave, they are entitled to disability benefits while they are out.

WORKERS COMPENSATION INSURANCE

If you employ permanent full or part-time employees in your home, you should consider obtaining a workers' compensation policy for your household employees. In the majority of states, workers' compensation coverage is mandatory and employers who fail to maintain the required insurance may be sued by the employee, and are liable for damages. In a minority of jurisdictions, workers' compensation coverage is elective, i.e., the employer may choose to reject the workers' compensation system. However, many states exclude household employees from their workers' compensation system.

If household employees are covered under state law, workers' compensation will be required through private insurance with an insurance company, or through a state insurance fund. It is advisable, however, to purchase a workers' compensation policy for your household employees even if your state does not require you to do so. In some jurisdictions, the employer is entitled to deduct a small contribution from the employee's salary to help defray the cost of the insurance.

The reader is advised to check the law of his or her own jurisdiction to determine whether workers compensation insurance is required for household employees.

Coverage

A workers compensation policy provides for medical care and rehabilitation for your employee if he or she is injured on the job. The policy also provides for lost wages if your employee is unable to work because of the injuries, as well as death benefits.

Filing the Claim

A claim for workers' compensation benefits is not a lawsuit. It is an application for benefits under an insurance policy for which a premium has been paid by the employer. Initiating a workers' compensation claim generally includes the following steps:

1. Notice to the Employer - In general, a work-related accidental injury is required to be reported within a certain number of days of its occurrence.

2. Filing the Claim - The claim for workers' compensation benefits must be filed with the insurance carrier. The employer is required to provide the employee with the necessary forms to be filed.

Making a false or fraudulent workers' compensation claim is a felony which, depending on the state, may subject the offender to imprisonment and/or a fine. It is also a felony for an employee to make a false statement for the purposes of obtaining compensation benefits, or for an employer to make a false statement for the purposes of denying compensation benefits.

A Directory of State Workers' Compensation Boards is set forth at Appendix 8.

A more detailed discussion of workers' compensation law can be found in this author's book entitled Workers' Compensation Law, published by Oceana Publishing Company.

EMPLOYEE PRACTICES LIABILITY (EPL) INSURANCE

As set forth in Chapter 1, a developing area of law affecting household employers is known as Employee Practices Liability (EPL). The household employer, particular if he or she has "deep pockets" and employs a large household staff, may be the target of a frivolous action brought by a disgruntled employee. Many of these lawsuits are based on discrimination in hiring, or wrongful termination.

You should speak with your insurance agent to find out whether you should add EPL coverage to your existing liability insurance, or if you should purchase a separate EPL insurance policy. If you purchase EPL coverage, you should make sure it covers all household employees, including part-time employees.

You should purchase enough coverage to limit your personal exposure to employment claims. Your EPL policy should also cover the cost of defending yourself against claims, including legal fees and costs. Once you purchase EPL coverage, you should keep the policy in force. EPL policies only provide coverage for claims brought while the policy is in effect unless there is a prior acts coverage endorsement on the policy.

UNEMPLOYMENT INSURANCE

Each state administers a separate unemployment insurance program within guidelines established by Federal law. Eligibility for unemployment insurance, benefit amounts and the length of time benefits are available are determined by the state law under which unemployment insurance claims are established.

In general, states provide unemployment benefits to eligible workers who are unemployed and who meet the eligibility requirements of state law, as set forth below:

1. The employee must meet the State requirements for wages earned or time worked during an established period of time referred to as a "base period."

2. The employee must be determined to be unemployed through no fault of their own, as determined under State law, and meet any other specific eligibility requirements of State law.

In order to avoid a financial catastrophe, a discharged or laid off employee should immediately apply for unemployment benefits to offset some of the lost income. There is usually a lag time of several weeks or more before the first check is received. Savings can quickly dissipate under such circumstances.

CHAPTER 5:
EMPLOYMENT TAXES

IN GENERAL

If you hire household help, you may be required to pay state employment taxes, such as state unemployment taxes, and federal employment taxes, which include social security, Medicare, and federal unemployment taxes, as well as withhold federal income tax. For tax purposes, it does not matter whether the household employee is full or part-time, or whether the employee is paid on a weekly or hourly basis. The penalties for not paying taxes on household employees include paying the back taxes and paying interest and penalty fines.

STATE EMPLOYMENT TAXES

State Unemployment Taxes

You must contact your state unemployment tax agency to find out whether you need to pay state unemployment tax for your household employee. You should also determine whether there are any other state employment taxes you must pay or collect, and whether you are required to carry workers' compensation insurance for your household employee.

A Directory of State Unemployment Tax Agencies is set forth at Appendix 9.

State Income Tax

Approximately 41 states require the employer to withhold state income tax.

State Disability Insurance

State disability insurance is also assessed on employees by some states, and is determined by setting a maximum withholding amount and/or a wage base.

FEDERAL EMPLOYMENT TAXES

Your obligation to pay federal employment taxes depends on cash wages paid to your household employee. Cash wages include wages you pay by check, money order, etc. Cash wages do not include the value of food, lodging, clothing, and other noncash items you give your household employee. However, cash you give your employee in place of these items is included in cash wages. If your employee was not paid the minimum cash wages set forth below, you do not need to pay any federal employment taxes, however, you may still need to pay state employment taxes.

Social Security and Medicare Taxes

The social security tax pays for old-age, survivors, and disability benefits for workers and their families. The Medicare tax pays for hospital insurance. You are responsible for payment of your employee's share of the taxes as well as your own. You can either withhold the employee's share from the employee's wages, or pay it from your own funds.

If you decide to pay your employee's social security and medicare taxes from your own funds, you must include the amount of those taxes paid on behalf of the household employee in the employee's wages for income tax purposes. However, they are not counted as social security and Medicare wages, or as federal unemployment (FUTA) wages.

If you pay cash wages of $1400 or more to any one household employee in 2005, regardless of when the wages were earned, all cash wages you pay to that employee are social security and Medicare wages. In addition, certain state disability plan payments that your household employee may receive are treated as social security and Medicare wages. However, any noncash wages you pay do not count as social security and Medicare wages.

Presently, social security and Medicare taxes are 15.3% of cash wages. Your share is 7.65%. Your employee's share of the taxes is 7.65%, which, as discussed above, you can choose to pay yourself and not withhold it.

You are permitted to withhold these taxes from your employees wages even if you are not sure whether he or she will earn the minimum amount. However, if you pay the employee less than $1,400 in cash wages in 2005, none of the wages you pay the employee are social security and Medicare wages, and neither you nor your employee will owe social security or Medicare tax on those wages. Thus, any amounts withheld from the employee's wages should be repaid to the employee.

Do not withhold any social security tax after your employee's social security wages for the year reach $90,000. However, you must continue to count the employee's cash wages as Medicare wages to figure the Medicare tax.

You do not have to count wages paid to your spouse, your child under the age of 21, your parent, or any employee under the age of 18 at any time in 2005, even if their wages totaled $1,400 or more during the year.

If the household service provided by the employee under age 18 is his or her principal occupation, you must count the wages. If the employee is a student, however, providing household services is not considered to be his or her principal occupation.

In addition, wages paid to your parent must be counted if both of the following conditions apply:

1. Your parent cares for your child who either:

    (i) is under the age of 18; or

    (ii) has a physical or mental condition that requires the personal care of an adult for at least 4 continuous weeks in a calendar quarter; and

2. Your marital status is one of the following:

    (i) You are divorced and have not remarried; or

    (ii) You are a widow or widower; or

    (iii) You are living with a spouse whose physical or mental condition prevents him or her from caring for your child for at least 4 continuous weeks in a calendar quarter.

Federal Unemployment Taxes

The federal unemployment tax is part of the federal and state program under the Federal Unemployment Tax Act (FUTA) that pays unemployment compensation to workers who lose their jobs. Like most employers, you may owe both the federal unemployment tax (FUTA tax) and a state unemployment tax. Or, you may owe only the FUTA tax or only the state unemployment tax.

As set forth above, to find out whether you will owe state unemployment tax, you must contact your states's unemployment tax agency.

If you pay total cash wages of $1,000 or more in any calendar quarter of 2005 to household employees, the first $7,000 of cash wages you pay in 2005 or 2006 is considered FUTA wages, and you must pay federal unemployment tax. If your employee's cash wages reach $7,000 during the year, do not figure the FUTA tax on any wages you pay the employee during the rest of the year.

Presently, the FUTA tax is 6.2% of your employee's FUTA wages. However, you may be able to take a credit of up to 5.4% against the FUTA tax, resulting in a net tax of 0.8% of cash wages. Your credit for 2005 is limited unless you pay all the required contributions for 2005 to your state unemployment fund by April 17, 2006. The credit you can take for any contributions for 2005 that you pay after April 17, 2006, is limited to 90% of the credit that would have been allowable if the contributions were paid by April 17, 2006.

You do not have to count wages paid to your spouse, your child under the age of 21, or your parent.

Federal Income Tax

You are not required to withhold federal income tax from wages you pay a household employee. You should withhold federal income tax only if your household employee asks you to withhold it, and you agree. At any time, however, either you or your employee may end the agreement by letting the other party know in writing.

If the employee chooses to have the federal income tax withheld, he or she must give you a completed Employee's Withholding Allowance Certificate (IRS Form W-4). If you agree to withhold federal income tax, you are responsible for paying it to the IRS. You must withhold the tax from each payment of wages based on the filing status and exemptions shown on the employee's Form W-4.

A sample Employee's Withholding Allowance Certificate (IRS Form W-4) is set forth at Appendix 10.

Earned Income Credit (EIC)

Certain employees can take the earned income credit (EIC) on their federal tax returns. This credit reduces their tax or allows them to receive a payment form the IRS. You may have to make advance payment of part of your household employee's EIC along with the employee's wages. You may also have to give your employee a notice about the EIC.

You must make advance EIC payments if your employee gives you a properly completed Earned Income Credit Advance Payment Certificate (IRS Form W-5). You can reduce the social security and Medicare taxes and withheld federal income tax you need to pay to the IRS by any advance EIC payments you make.

A sample Earned Income Credit Advance Payment Certificate (IRS Form W-5) is set forth at Appendix 11.

MAKING THE TAX PAYMENTS

You must use IRS Schedule H (Household Employment Taxes) to figure your total household employment taxes, including social security, Medicare, FUTA and withheld federal income taxes. You must add these household employment taxes to your income tax when you file your federal income tax return, and attach Schedule H to your IRS Form 1040. If you are not filing a tax return, e.g., because your income is under the amount that requires you to file, you must nevertheless file Schedule H along with your payment for the total household employment taxes due.

A sample IRS Schedule H is set forth at Appendix 12.

OBTAINING AN EMPLOYER IDENTIFICATION NUMBER (EIN)

When you file your tax forms, you must include your employer identification number (EIN). An EIN is a 9-digit number issued by the IRS. It is not the same as your social security number. If you do not have an EIN, you can apply for one by completing an Application for Employer Identification Number (IRS Form SS-4). You can apply by telephone, mail, fax or internet, depending on how soon you need your EIN.

A sample Application for Employer Identification Number (IRS Form SS-4) is set forth at Appendix 13.

Telephone

You can obtain an EIN instantly by applying over the telephone (1-800-829-4933) between the hours of 7:00a.m. to 10:00p.m.

Internet

You can also obtain your EIN immediately online through the IRS website at www.irs.gov/businesses.

FAX-TIN

Under the IRS Fax-TIN program, you can receive your EIN by fax within 4 business days. Fax-TIN is available 24 hours a day, 7 days a week.

Mail

You can also apply for your EIN by mail, which takes about 4 weeks to process.

A Directory of IRS addresses and Fax-TIN numbers, by state, is set forth at Appendix 14.

FILING YOUR EMPLOYEE'S WAGE AND TAX STATEMENT (IRS FORM W-2)

You must file a separate Wage and Tax Statement (IRS Form W-2) for each household employee to whom you pay either of the following wages during the year:

1. Social security and Medicare wages of $1,400 or more; or
2. Wages from which you withhold federal income tax.

IRS Form W-2 has multiple parts. You must complete the form and give copies B, C and 2 to your employee by January 31st of the following year, and send Copy A to the Social Security Administration (SSA) by February 28th of the following year. Employers who file their forms electronically at the SSA website (www.socialsecurity.gov/employer) do not have to file until March 31st of the following year.

A sample Employee Wage and Tax Statement (IRS Form W-2) is set forth at Appendix 15.

RECORDKEEPING

When you employ household workers, it is important to keep accurate records to demonstrate your compliance with all of the laws concerning hiring household help, if you are called upon to do so. You must keep copies of all of the forms you file with the Internal Revenue Service, and all records to support the information you enter on the forms you file.

In addition, on each payday, you should record the date and amounts of all of the following items:

1. Your employee's cash and noncash wages.
2. Any employee social security tax you withhold or agree to pay for your employee.
3. Any employee Medicare tax you withhold or agree to pay for your employee.
4. Any federal income tax you withhold.
5. Any advance earned income credit payments you make.
6. Any state employment taxes you withhold.

You must also keep a record of your employee's name and social security number exactly as they appear on his or her social security card if you pay the employee either of the following:

1. Social security and Medicare wages of $1,400 or more.
2. Wages from which you withhold federal income tax.

You must ask for your employee's social security number no later than the first day on which you pay wages. Ideally, you should obtain the number when the employee is hired, and verify it against his or her social security card.

You should keep your employment tax records for at least 4 years after the due date of the return on which you report the taxes, or the date the taxes were paid, whichever is later.

CHAPTER 6: EMPLOYEE OR INDEPENDENT CONTRACTOR?

IN GENERAL

If you hire someone to work for you as your employee, you must comply with all legal requirements discussed in this almanac. However, if the individual you hire is an independent contractor, your legal obligations are minimal. An individual who serves in the capacity of an independent contractor may not have the recourse under the law afforded an employee. Thus, whether a worker is deemed an employee or an independent contractor is important to both the employer and the worker.

INDEPENDENT CONTRACTOR DEFINED

An independent contractor is generally defined as a person who contracts with another to perform services for that person, but who is not controlled by the other nor subject to the other's right to control with respect to the performance. Thus, if an employer has the right to control when the work is done, where it is done, and how it is done, the worker would likely be considered an employee. If the worker controls these conditions, he or she would more likely be considered an independent contractor, particularly if the worker provides his or her own equipment.

EMPLOYER-EMPLOYEE RELATIONSHIP COMPARED

An employer-employee relationship is very different from the relationship between an employer who hires an independent contractor. Employees have a special legal status. As shown below, employers have certain responsibilities towards employees that they do not have when they engage the services of an independent contractor.

1. When you hire an employee, you pay the employee a salary and you are legally required to make certain payroll deductions for such items as taxes and social security. The independent contractor receives a gross income check with no deductions for taxes, social security or other customary deductions from an employee's paycheck. Instead, the independent contractor estimates his or her taxes and pays those estimated amounts to the IRS at certain yearly intervals. In addition, an independent contractor must pay self-employment taxes, which includes their Social Security and Medicare taxes.

2. An employee usually has no deductions for business expenses since all of the employee's business is conducted on your premises. An independent contractor is entitled to certain tax deductions that an employee does not have. For example, business expenses such as equipment, space rental, utilities, transportation, and entertainment expenses are deductible.

3. An employee follows your instructions as to when, where and how to perform the services for which he or she is hired. An independent contractor has the freedom to make his or her own decisions, e.g. regarding what jobs to take, how much to charge, and when to work, etc.

4. An employee may be entitled to certain benefits, such as unemployment compensation, workers' compensation, health insurance, vacation pay, etc. An independent contractor is not entitled to any benefits in connection with the work performed.

IRS RECLASSIFICATION

For the above reasons, an employer may wish to hire an individual and classify them as an independent contractor. Nevertheless, because the tax treatment of an employee is very different from that of an independent contractor, the Internal Revenue Service may attempt to reclassify independent contractors as employees. For example, the Internal Revenue Service has already determined that all nannies are employees.

IRS reclassification can be very costly to the employer, in that the employer would then be liable for withholding tax, Social Security contributions, and unemployment tax withholding, as well as penalties and interest on those amounts due.

To determine whether or not an individual is likely to be reclassified by the IRS from independent contractor status to employee status, courts have established 20 common law factors which must be taken into account. The factors that may indicate employee status are the following:

1. Whether the worker is required to comply with the employer's instruction as to when, where, and how he or she is to work.

2. Whether the employer trains the worker to perform in a particular manner.

3. Whether the employer requires the worker to render personal service.

4. Whether the employer hires, supervises, and pays the worker's staff.

5. Whether the employer and the worker have a continuing relationship.

6. Whether the worker is required to work during hours set by the employer.

7. Whether the worker is required to work full-time for the employer.

8. Whether the worker performs the work on the employer's premises, particularly if the work could be performed elsewhere.

9. Whether the employer requires the worker to perform services in a particular order or sequence.

10. Whether the worker is required to submit oral or written reports to the employer.

11. Whether payment is made by the hour, week, or month as opposed to a per job basis.

12. Whether the employer pays the worker's business or traveling expenses.

13. Whether the employer furnishes the tools and materials to the worker.

14. Whether there is a lack of investment by the worker in the facilities required to perform services and dependence on the employer for such facilities.

15. Whether the worker realizes a profit or loss as a result of the services rendered.

16. Whether the worker works exclusively for the employer.

17. Whether the worker makes his or her services available to the general public.

18. Whether the employer maintains the right to discharge the worker, which would not be possible if the individual were an independent contractor working under a contract.

19. Whether the worker has the right to end the relationship with the employer without continuing to render services. An independent contractor would be bound by the contract to finish the job.

20. Whether the worker's services are integral to the business operations.

Based on the above factors, the vast majority of household workers would likely be classified as employees, not independent contractors. Following are some examples:

EXAMPLE: You hire a housekeeper to clean your home and cook meals five days a week, with certain hours, e.g., 9am - 6pm. You prepare the menus and instruct the housekeeper as to how you would like the house kept, e.g., vacuum floors every day, wash windows once per week, etc. You also provide all of the supplies, food for the meals, etc. Under this scenario, your housekeeper would be considered an employee.

EXAMPLE: You hire a worker to maintain your pool during the summer months, such as cleaning the pool, checking the chemical levels, etc. The worker performs the same maintenance procedure for a number of other customers. He uses his own equipment and supplies, and arranges his visits at times that fit into his schedule. He bills you on a monthly basis. Under this scenario, the pool maintenance worker would be considered an independent contractor.

CHAPTER 7: HIRING A CONTRACTOR

IN GENERAL

Although most licensed contractors are reputable and competent individuals, a consumer who does not plan carefully and research the prospective contractor is more likely to be the victim of an unscrupulous, unlicensed and inferior worker. In fact, consumers nationwide routinely register complaints over dissatisfaction with their home improvement contractors, including unexpected costs and unreasonable delays. You can avoid these problems by taking the time to properly plan your home improvement project.

PLANNING YOUR PROJECT

Whether the job entails installing kitchen cabinets, replacing your roof, or adding an addition to your home, you must carefully plan each and every phase of the project. Determine exactly what you want the contractor to accomplish, and the amount of money you are willing to spend to get the job done. For large remodeling jobs, you may want to engage an architect to lay out precise and detailed plans for the contractor.

SELECTING THE RIGHT CONTRACTOR FOR THE JOB

There are many different types of contractors, each with their own specialty. A general building contractor usually oversees large projects, and coordinates the various subcontractors needed to complete the job. For example, if you are remodeling your kitchen, you will likely need plumbing, electrical and carpentry work. In that case, a general contractor should be hired. One contract would be negotiated with the general contractor instead of a contract with each individual subcontractor.

However, if it is a specific task you need done, you only need to hire the contractor who specializes in that particular area. For example, if your roof needs repair, you need a roofing contractor.

A description of contractor classifications is set forth at Appendix 16.

CHECKING CREDENTIALS

Before you hire a contractor, you should check his or her credentials. You should only hire a contractor who is licensed in the class of work you need performed. Don't become a victim of shoddy work just because you are quoted a really low bid price for the job. It is risky to hire an unlicensed contractor. He or she may not have proper insurance, which may expose you to financial liability in case a worker sustains an injury on your premises. In addition, the unlicensed contractor may not have the expertise to properly complete the job, in which case you will likely end up hiring a licensed contractor to correct the problems, creating additional expense.

The contractor's license must be up-to-date and in good standing. In order to check the contractor's license, you must contact your state's licensing board. You should also ask the contractor to show you his or her license and another form of identification. Make sure the name on the license matches the name on the contractor's identification. It is illegal for one contractor to use another contractor's license.

Verify the contractor's business address and telephone number. Beware of the contractor who can only be reached by cell phone. If the contractor does not have a permanent business address and telephone number, it may be difficult locating him or her if a problem arises after the job is done and payment has been made.

Ask the prospective contractor for references. Call the references and ask them what they thought about the contractor's work habits. For example, ask whether the work was performed according to schedule. If possible, visit other similar projects that the contractor is working on, or has recently finished.

You should also check your local Better Business Bureau and consumer protection agency to find out whether there have been any complaints made against the contractor. You can also find out if there are any lawsuits pending against the contractor by visiting the clerk's office in your local court.

Common warning signs that may indicate future problems with a particular contractor include:

1. You can't verify the name, address and telephone number or credentials of the contractor.

2. The contractor gives you a toll-free phone number and a post office box as his or her address.

3. The salesperson tries to pressure you into signing a contract by using scare tactics, intimidation or threats.

4. The company or salesperson says your home will be used for advertising purposes, and that you will be given a special low price.

5. The contractor tells you this is a special price available only if you sign the contract today.

6. The contractor doesn't comply with your request for references, or the references have some reservations about the contractor's work.

7. You are unable to verify that the contractor is licensed, insured, and/or bonded when required.

8. You are asked to pay for the entire job in advance, or to pay cash to a salesperson instead of writing a check or money order to the company itself.

9. You are asked to sign a completion certificate for the job by appeal, threat, or trick, before the job is properly completed.

INSURANCE COVERAGE

You must find out whether the contractor is insured against claims covering workers' compensation, property damage, and personal liability in case of accidents. You should ask to see a copy of the certificate of insurance, or ask for the name of the contractor's insurance carrier and agency to verify that the contractor has the appropriate insurance coverage.

Licensed contractors are not usually required to carry general liability insurance. However, they should disclose to you whether or not they carry such insurance. If they are uninsured, you must ask them how they intend to cover any losses that would ordinarily be covered by insurance. This matter should be covered in your contract. If your property is damaged and the contractor is uninsured, you may end up having to pay for the damages out of your own pocket, or through your homeowner's insurance policy.

Most states require contractors who have employees to have workers' compensation coverage. This protects you if a worker is injured while working on your property. If the contractor does not have workers' compensation coverage, you may have to go through your homeowner's insurance policy to pay medical bills and compensate the injured worker.

Workers' compensation is discussed more fully in Chapter 4 of this almanac, and in this author's legal almanac entitled Workers' Compensation Law, published by Oceana Publishing Company.

The reader is advised to check the law of his or her own jurisdiction concerning insurance requirements for contractors.

BONDS

Bonds are used to protect the consumer against substandard work that does not comply with the local building codes. Bonds do not guarantee the contractor is competent. Lenders usually require contractors to secure bonds for large projects.

BIDS

A bid is an offer to do work at a specific price. It is advisable to get at least three written bids using identical plans and specifications so you can compare prices and contractors. However, do not automatically accept the lowest bid. Discuss the bids in detail with each contractor, and make sure you understand the reasons for any price variations.

Sometimes, paying a higher price may be worth it, particularly if the materials to be used are of higher quality, the work is more extensive, or the contractor is more experienced and has a better track record. Be particularly wary of any bid that is substantially lower than the other two. The contractor may have made a mistake in computation, or may not be including all of the work needed to complete the job.

CHAPTER 8: THE HOME IMPROVEMENT CONTRACT

IN GENERAL

Do not sign anything until you understand what you are signing and agree to all of the terms. This is the first and most important rule when entering into a contract with a contractor for any type of home improvement. You do not want to unintentionally give the contractor the right to start work by signing something that may be legally binding. The contract is the document that binds you and the contractor to the project.

GET IT ALL IN WRITING

Since a written contract protects both you and the contractor, all agreements should be put in writing. The contract should include everything you have agreed upon, and detail the extent of work to be done. Spell out exactly what the contractor will and will not do. Be as specific as possible. If the contractor made any oral promises, make sure they are put in writing and included in the contract.

After you have a signed contract, and even after work has already begun, you or your contractor may have suggestions that will alter your original plan. Any changes should also be put in writing, in a "change order," and signed by both parties.

Never sign a blank or partially blank contract. Obtain a copy of the contract as soon as you sign it, and keep it for your records. Both you and the contractor are bound by everything set down in the contract, so read it carefully before you sign. As discussed below, you should make sure the financial terms are set forth in the contract, including the total price, when payments will be made, and whether there is a cancellation penalty. Generally, the contractor will require a down payment at the time the contract is signed.

If you have any questions or don't understand something contained in the contract, ask for clarification before you sign. For larger, more expensive jobs, you may want to consult an attorney to make sure you are adequately protected under the contract.

A sample homeowner/contractor contract for services is set forth at Appendix 17.

MATERIALS

The contract should specify all materials to be used, including the quality, quantity, weight, color, size, or brand name, etc. For example, if the contract calls for the installation of a kitchen sink, the contract should specify the type of sink, the manufacturer, etc. The contract should not simply state "Contractor agrees to install kitchen sink." Also, make sure your contract includes everything you feel is important to the job, including cleanup and removal of debris and materials. In addition, your contract should state that all work will be performed in accordance with applicable building codes.

PAYMENT SCHEDULE

Your contract should contain the payment schedule. You should never pay the contractor the total contract price before the job is completed. If so, you remove the incentive to complete the job as scheduled. The payment schedule should be based on the contractor's performance. Make sure the contract provides for a "retention"—i.e., a percentage of the total price which you retain until the job is completed.

Also, never sign a completion certificate until all the work specified in the contract has been properly completed. If a lender is involved, the will . . . usually require a signed completion certificate before they will release the last payment.

PROJECT START/END DATE

Your contract should specify an approximate starting date and completion date for your project. The contract should not simply state the amount of time it takes to complete the job, e.g., 7 days. If the project must be completed by a certain date, you should put a "time is of the essence" clause in the contract, and include a provision awarding you monetary penalties to be assessed for each day the contractor goes beyond the stated completion date.

CANCELLATION PERIOD

The law generally requires a contractor to give the consumer written notice of their right to cancel a contract within a certain number of days—e.g., three business days—of signing the contract. You should carefully review the contract during this cancellation period. If something doesn't look right, do not hesitate to cancel the contract.

Although you should call the contractor and tell him or her you want to cancel, you should also put your cancellation notice in writing and mail it to the contractor before the cancellation period ends. You should send the cancellation notice by certified mail/return receipt requested, so that you have proof of the date the notice was mailed, and proof of its receipt by the contractor.

The reader is advised to check the law of his or her jurisdiction concerning specific cancellation provisions.

WARRANTIES

If the contractor offers a warranty on labor and/or materials, make sure you get the warranty in writing. The warranty should specify the work that is covered, and the duration of the warranty. If there are any written warranties from the manufacturers of the materials used in the project, you should request a copy of those warranties as well.

KEEPING A PROJECT FILE

You should keep a file for all of the papers related to the project, including:

1. The contract and any change orders.
2. Plans and specifications.
3. Bills and invoices.
4. Canceled checks.
5. Lien releases from subcontractors and material suppliers.
6. Letters, notes, and correspondence with your contractor.
7. Pictures of the job in progress.
8. A record of each subcontractor who works on your project, the work performed, and length of time on the job.
9. A list of materials delivered, including the name of the material supplier, and date and description of materials delivered.
10. Warranties.

BUILDING CODES AND PERMITS

As a general rule, a building permit is required whenever structural work is involved or when the basic living area of a home is to be changed. For example, if you have a home with an unfinished basement, and would like to finish off a portion of it for a family room, you would need a building permit in most jurisdictions because you have increased the "livable" space in your house. In some cases, separate permits for electrical, heating and plumbing may be required.

The contractor should obtain the necessary building permits. This should be spelled out in your contract, otherwise you may be held legally responsible for failure to obtain the required permits. In general, homeowners are not required to sign building permits unless they are performing their own work personally.

A frequent practice of unlicensed contractors is to have the homeowner secure an "owner-builder" building permit, which infers that the property owner is providing his or her own labor and materials personally. The term "owner/builder" refers to a homeowner who acts as the general contractor. The owner/builder assumes responsibility for the overall job, which may include such things as state and federal taxes, workers' compensation, building permits and other legal liabilities. Unless you are very experienced in construction, it is best to leave these matters to a licensed contractor.

Insofar as zoning regulations differ from place to place, your contractor should check with the zoning authorities to determine what permits or permission you need to proceed with any alterations or additions to your home. The permit for your job, along with the project plans and specifications, must be posted on the job before the work begins.

INSPECTIONS

The building department will send an inspector out to examine the work when it has reached a certain stage, and again when it is completed, to make sure it complies with various codes and regulations. The contractor is responsible for arranging for these inspections. If possible, you should be present when inspections are made, and ask questions.

When a project is completed, the building department will make a final inspection, so don't make the final payment to the contractor until the building department inspector has approved the job.

MECHANIC'S LIENS

A mechanic's lien is a lien filed against a customer's home by an individual who furnished labor or materials, and who has not been paid. If you undertake a large remodeling project, you will probably hire a general contractor, who in turn will hire a number of subcontractors and use a variety of suppliers for the materials. In that case, you should protect yourself from liens against your home in case the general contractor fails to pay the subcontractors or the suppliers.

Even if you pay the general contractor, you still run the risk of having a mechanic's lien filed against your home if the general contractor fails to pay any of the subcontractors or suppliers. That means that you could end up having to pay a bill twice in order to avoid foreclosure proceedings against your home by the party trying to enforce the lien. In addition, the lien creates a "cloud" on the title of the property, which can prevent you from refinancing or selling your home. There are several ways to handle this situation, as set forth below.

Waiver and Release

You can protect yourself from unwarranted liens by identifying subcontractors and materials suppliers in your written contract, and getting waiver and release forms signed by the subcontractors and suppliers.

Conditional Waiver and Release Upon Progress Payment

A Conditional Waiver and Release Upon Progress Payment form is used when the subcontractor or supplier has not been paid yet, but will be paid out of the monies paid at various times as the project progresses, pursuant to the contract, before the final payment is made. However, the conditional waiver and release is only effective if the claimant is actually paid. Evidence of payment may be demonstrated by:

1. The claimant's endorsement on a single check or a joint payee check which has been paid by the bank upon which it was drawn; or

2. Written acknowledgment of payment given by the claimant.

A sample Conditional Waiver and Release Upon Progress Payment Form is set forth at Appendix 18.

Unconditional Waiver and Release Upon Progress Payment

An Unconditional Waiver and Release Upon Progress Payment form is used when the claimant is required to execute a waiver and release in exchange for or in order to induce payment of a progress payment, and the claimant asserts in the waiver that he or she has in fact been paid the progress payment.

A sample Unconditional Waiver and Release Upon Progress Payment Form is set forth at Appendix 19.

Conditional Waiver and Release Upon Final Payment

A Conditional Waiver and Release Upon Final Payment form is used when the claimant is required to execute a waiver and release in exchange for or in order to induce the payment of a final payment and the claimant has not been paid. This release is only binding if there is evidence of payment to the claimant, as set forth above.

A sample Conditional Waiver and Release Upon Final Payment Form is set forth at Appendix 20.

Unconditional Waiver and Release Upon Final Payment

An Unconditional Waiver and Release Upon Final Payment form is used when the claimant is required to execute a waiver and release in exchange for, or in order to induce payment of, a final payment and the claimant asserts in the waiver he or she has in fact been paid the final payment.

A sample Unconditional Waiver and Release Upon Final Payment Form is set forth at Appendix 21.

Joint Checks

You can pay the general contractor with checks that are jointly made out to the general contractor and the subcontractor or supplier as each bill is presented to you. That will ensure that the subcontractors and suppliers are paid, and your checks will serve as proof that you paid them.

Notice of Completion

In some states, if you file a Notice of Completion with the clerk's office after the work on your project is completed, the amount of time a contractor, subcontractor or materials supplier can file a mechanic's lien against your home is reduced.

The reader is advised to check the law of his or her jurisdiction to find out whether this method is available.

If you are financing the project, your lender may require the general contractor, subcontractors and suppliers to verify that they have been paid as a prerequisite to releasing funds upon the completion of various phases of the project.

Invalid Liens

It is possible that a lien filed against your property is not valid. For example, a lien is not valid if the work was not completed. In addition, many lien claims are invalid because the contractor, subcontractor or supplier has failed to meet the required timelines for filing the claim.

many lien claims are invalid because the contractor, subcontractor or supplier has failed to meet the required timelines for filing the claim.

If you are faced with this situation, you can try and get the lien claimant to release the lien voluntarily. Send the lien claimant a written request by certified mail/return receipt requested. Keep a copy of your letter and the certified mail receipt as proof of your request. In your letter, explain the reasons why you believe the lien is invalid, and request the lien claimant to remove the lien.

You can include a mild threat in your letter, advising the lien claimant that his or her failure to remove the invalid lien will cause you to have to hire an attorney to go to court, and that he or she may be held liable for attorneys fees and costs. Such a letter may be enough incentive to get the lien claimant to release the lien. If the letter doesn't work, and the lien claimant refuses to remove the invalid lien, you may have to hire an attorney to go to court and get the lien released.

DISPUTE RESOLUTION

If you are displeased with some aspect of the work, either during construction or following completion, you should contact the contractor. Most reputable contractors will make any necessary corrections. Follow up any verbal conversation with written confirmation.

If the contractor refuses to make the corrections, you should file a complaint with your state's licensing board and department of consumer protection. Your complaint will be investigated. You may also want to consult an attorney in case litigation is necessary.

Arbitration and Mediation

If you include an arbitration or mediation clause in your contract, both parties give up your right to have any disputes litigated in court by a judge or jury. Instead, a neutral arbitrator or mediator will listen to both sides, and make a decision.

A sample contract arbitration clause and mediation clause is set forth at Appendix 22.

Arbitration and mediation is discussed more fully in this author's legal almanac entitled The Law of Dispute Resolution, published by Oceana Publishing Company.

APPENDIX 1: HOUSEHOLD EMPLOYER/EMPLOYEE EMPLOYMENT AGREEMENT

THIS AGREEMENT is entered into between [Name of Employer] ("Employer"), and [Name of Employee] ("Employee") as of the date signed below.

Employer agrees to hire and employ Employee, and Employee agrees to be so employed as a [specify position/title] on the following terms:

1. Employee's name, address and telephone are as follows: [Set forth identifying information].

2. Employee's social security number is: [Set forth social security number].

3. Employee agrees to perform the following services: [Specify duties] pursuant to the attached job description [Attach copy of job description].

4. Employee agrees to faithfully perform the duties assigned to him to the best of Employee's ability, and to devote Employee's full and undivided time to the transaction of Employer's business.

5. In consideration of Employee's service, Employer agrees to pay Employee compensation at the rate of [Dollar Amount ($xxx)] Dollars per [specify manner in which salary is to be paid, e.g. weekly, monthly, etc.] and Employee's reasonable and necessary traveling expenses incurred in Employer's business while away from the usual place of business.

6. Employee's schedule is as follows: [Specify days and hours]

7. Employee shall be entitled to the following benefits: [Specify benefits, e.g., 10 paid vacation days per year]

8. This contract shall be in effect from [Insert Effective Date] until [Insert Termination Date or other method of termination, e.g., . . . until either party gives 30 days notice to the other that they wish to terminate the agreement].

9. Notwithstanding the foregoing paragraph, the following actions will result in immediate termination without notice [specify actions that will result in termination, e.g. theft, lateness, etc.].

10. Upon the termination of this agreement as set forth above, the payment to Employee of salary earned to the date of such termination shall be in full satisfaction of all claims against Employer under this agreement.

DATED:

BY: ____

[SIGNATURE LINE - EMPLOYER]

BY: ____

[SIGNATURE LINE - EMPLOYEE]

APPENDIX 2: STATE MINIMUM AGE REQUIREMENT FOR EMPLOYMENT

STATE	MINIMUM AGE REQUIREMENT
Alabama	16
Alaska	16
Arizona	14
Arkansas	14
California	15
Colorado	16
Connecticut	16
Delaware	14
District of Columbia	14
Florida	14
Georgia	16
Hawaii	16
Idaho	14
Illinois	16
Indiana	14
Iowa	16
Kansas	14
Kentucky	16
Louisiana	16
Maine	16
Maryland	16
Massachusetts	16
Michigan	14

STATE	MINIMUM AGE REQUIREMENT
Minnesota	14
Mississippi	14
Missouri	14
Montana	16
Nebraska	14
Nevada	14
New Hampshire	16
New Jersey	16
New Mexico	14
New York	16
North Carolina	16
North Dakota	14
Ohio	16
Oklahoma	14
Oregon	14
Pennsylvania	16
Rhode Island	16
South Carolina	16
South Dakota	14
Tennessee	14
Texas	15
Utah	16
Vermont	14
Virginia	16
Washington	14
West Virginia	16
Wisconsin	16
Wyoming	16

APPENDIX 3:
TABLE OF EMPLOYMENT CERTIFICATE ISSUANCE PRACTICES UNDER STATE CHILD LABOR LAWS

State	Age Requirement	Issuing Authority	Comments
Alabama	Under 18	School	Required for minors age 18 for work in mines
Alaska	Under 17	Labor Department	Required for minors under 19 if employer licensed to sell alcohol; In addition to individual certificates, employers may obtain advance approval for a specific job consisting of listed duties permitting them to hire minors of at least 14 years of age without prior individual approval.
Arizona	Not issued	N/A	None
Arkansas	Under 16	Labor Department	None

State	Age Requirement	Issuing Authority	Comments
California	Under 18	School	Required for minors enrolled in school
Colorado	Under 16	School	Required for work during school hours
Connecticut	Under 16	School	None
Delaware	Under 18	Labor Department or School	None
District of Columbia	Under 18	School	None
Florida	Not issued	N/A	Employers of any minor must obtain and keep on record proof of the child's age
Georgia	Under 18	School	None
Hawaii	Under 16	Labor Department	None
Idaho	Not issued	N/A	None
Illinois	Under 16	School	None
Indiana	Under 18	School	None
Iowa	Under 16	School	Employment certificates are issued by both the Division of Job Service of the Department of Employment Services and the schools
Kansas	Under 16	Labor Department	Required for minors not enrolled in secondary school
Kentucky	Not issued	N/A	Employers of children under age 18 must maintain a proof of age
Louisiana	Under 18	School	None
Maine	Under 16	Labor Department	None

State	Age Requirement	Issuing Authority	Comments
Maryland	Under 18	Labor Department or School	None
Massachusetts	Under 16	School	None
Michigan	Under 18	School	None
Minnesota	Under 16	N/A	Required for work during school hours; employers of any minor must obtain and keep on record proof of the child's age
Mississippi	Under 16	School	Required for work in mills, canneries, workshops, and factories
Missouri	Under 16	School	None
Montana	Not issued	N/A	No certificate is required at any age in nonhazardous employment and since minors under 16 may not work in hazardous employment, certificates for this age category are not deemed necessary
Nebraska	Under 16	School	None
Nevada	Under 14	School	Certificates issued with written permission of district judge
New Hampshire	Under 16	Labor Department	Work certificates are not required for 16- and 17-year-olds but employers must maintain on file a signed, written document from the youth's parent or legal guardian authorizing the employment
New Jersey	Under 18	School	None
New Mexico	Under 16	Labor Department or School	None
New York	Under 18	School	None

State	Age Requirement	Issuing Authority	Comments
North Carolina	Under 18	Labor Department or County Director of Social Services	None
North Dakota	Under 16	Labor Department	None
Ohio	Under 16	School	Required for minors aged 16 and 17 during school term; proof of age is required for minors 16 and 17 years of age for employment during school vacation; 16- and 17-year-old minors are not required to provide a certificate to be employed at a seasonal amusement or recreational establishment
Oklahoma	Under 16	School	None
Oregon	Not issued	N/A	Minors aged 14 – 17 are not required to obtain work permits but employers are required to apply for annual certificates to employ such minors
Pennsylvania	Under 18	School	None
Rhode Island	Under 16	School	None
South Carolina	Not issued	N/A	None
South Dakota	Not issued	N/A	None
Tennessee	Not issued	N/A	Employers of minors under age 18 must obtain and keep on file proof of the minor's age. A birth certificate, passport, driver's license, State issued identification, or parent's oath as to the minor's age are acceptable proofs of age under the child labor law
Texas	Not issued	N/A	None
Utah	Not issued	N/A	None

State	Age Requirement	Issuing Authority	Comments
Vermont	Under 16	Labor Department	Required for work during school hours
Virginia	Under 16	School	None
Washington	Under 18	Labor Department	None
West Virginia	Under 16	School	None
Wisconsin	Under 18	Labor Department	Certificates are issued by volunteer permit officers who are representatives of the Department of Industry, Labor, and Human Relations; many of the permit officers are school officials, but other public employees such as municipal and county employees and court officials are also included
Wyoming	Not issued	N/A	Employers of children under age 16 must maintain a proof of age

Source: U.S. Department of Labor

APPENDIX 4: DIRECTORY OF STATE DEPARTMENTS OF LABOR

STATE	ADDRESS	TELEPHONE	FAX	WEBSITE
ALABAMA	Department of Labor P.O. Box 303500 Montgomery, AL 36130-3500	334-242-3460	334-240-3417	www.dir.state.al.us
ALASKA	Department of Labor P.O. Box 21149 Juneau, AK 99802-1149	907-465-2700	907-465-2784	www.labor.state.ak.us
ARIZONA	State Labor Department P.O. Box 19070 Phoenix, AZ 85005-9070	602-542-4411	602-542-3104	www.ica.state.az.us
ARKANSAS	Department of Labor 10421 West Markham Little Rock, AR 72205	501-682-4541	501-682-4535	www.state.ar.us/labor

STATE	ADDRESS	TELEPHONE	FAX	WEBSITE
CALIFORNIA	Department of Industrial Relations 455 Golden Gate Ave., 10th Floor San Francisco, CA 94102	415-703-5050	415-703-5059	www.dir.ca.gov
COLORADO	Department of Labor and Employment 2 Park Central, Suite 400 Denver, CO 80202-2117	303-620-4701	303-620-4714	http://cdle.state.co.us
CONNECTICUT	Labor Department 200 Folly Brook Boulevard Wethersfield, CT 06109-1114	860-263-6505	860-263-6529	www.ctdol.state.ct.us
DELAWARE	Department of Labor 4425 N. Market Street, 4th Floor Wilmington, DE 19802	302-761-8000	302-761-6621	www.delawareworks.com
DISTRICT OF COLUMBIA	Department of Employment Services Employment Security Building 500 "C" Street NW, Suite 600 Washington, D.C. 20001	202-724-7100	202-724-5683	does.ci.washington.dc.us
FLORIDA	Department of Labor and Employment Security 2012 Capitol Circle S.E., Hartman Building, Suite 303 Tallahassee, FL 32399-2152	850-922-7021	904-488-8930	www.state.fl.us/dles
GEORGIA	Department of Labor Sussex Place - Room 600 148 International Blvd. N.E. Atlanta, GA 30303	404-656-3011	404-656-2683	www.dol.state.ga.us

STATE	ADDRESS	TELEPHONE	FAX	WEBSITE
HAWAII	Department of Labor and Industrial Relations 830 Punchbowl Street, Room 321 Honolulu, HI 96813	808-586-8844	808-586-9099	www.dlir.state.hi.us
IDAHO	Department of Labor 317 W. Main Street Boise, ID 83735-0001	208-334-6110	208-334-6430	www.labor.state.id.us
ILLINOIS	Department of Labor 160 N. LaSalle Street, 13th Floor, Suite C-1300 Chicago, IL 60601	312-793-1808	312-793-5257	www.state.il.us/agency/idol
INDIANA	Department of Labor 402 West Washington Street, Room W195 Indianapolis, IN 46204-2739	317-232-2378	317-233-5381	www.state.in.us/labor
IOWA	Workforce Development 1000 East Grand Avenue Des Moines, IA 50319-0209	515-281-5365	515-281-4698	www.state.ia.us/iwd
KANSAS	Department of Human Resources 401 S.W. Topeka Boulevard Topeka, KS 66603	785-296-7474	785-368-6294	www.hr.state.ks.us
KENTUCKY	Labor Cabinet 1047 U.S. Hwy. 127 South, Suite 4 Frankfort, KY 40601	502-564-3070	502-564-5387	www.state.ky.us/agencies/labor/labrhome.htm

STATE	ADDRESS	TELEPHONE	FAX	WEBSITE
LOUISIANA	Department of Labor P.O. Box 94094 Baton Rouge, LA 70804-9094	225-342-3011	225-342-3778	www.ldol.state.la.us
MAINE	Department of Labor State House Station #45 Augusta, ME 04333	207-624-6400	207-624-6449	http://janus.state.me.us/labor
MARYLAND	Department of Labor Licensing and Regulation, 1100 Eutaw St., 6th Floor Baltimore, MD 21201	410-767-2999	410-767-2986	www.dllr.state.md.us
MASSACHUSETTS	Department of Labor & Work Force Development 1 Ashburton Place, Rm 2112 Boston, MA 02108	617-727-6573	617-727-1090	www.state.ma.us
MICHIGAN	Department of Consumer & Industry Services P.O. Box 30004 Lansing, MI 48909	517-373-3034	517-373-2129	www.cis.state.mi.us/bsr/divisions/whd/home.htm
MINNESOTA	Department of Labor and Industry 443 Lafayette Road St. Paul, MN 55155	651-296-2342	651-282-5405	www.doli.state.mn.us
MISSISSIPPI	Employment Security Commission, P.O. Box 1699 Jackson, MS 39215-1699	601-961-7400	601-961-7406	www.mesc.state.ms.us

STATE	ADDRESS	TELEPHONE	FAX	WEBSITE
MISSOURI	Department of Labor & Industrial Relations P.O. Box 504 Jefferson City, MO 65102	573-751-9691	573-751-4135	www.solir.state.mo.us
MONTANA	Department of Labor and Industry P.O. Box 1728 Helena, MT 59624-1728	406-444-9091	406-444-1394	http://dli.state.mt.us
NEBRASKA	Department of Labor 550 South 16th Street Box 94600 Lincoln, NE 68509-4600	402-471-9792	402-471-2318	www.dol.state.ne.us
NEVADA	Business and Industry 555 E. Washington Avenue, Suite 4100 Las Vegas, NV 89101	702-486-2650	702-486-2660	www.state.nv.us/labor
NEW HAMPSHIRE	Department of Labor 95 Pleasant Street Concord, NH 03301	603-271-3171	603-271-6852	www.state.nh.us/dol
NEW JERSEY	Department of Labor John Fitch Plaza, 13th Floor P.O. Box CN 110 Trenton, NJ 08625-0110	609-292-2323	609-633-9271	www.state.nj.us/labor
NEW MEXICO	Department of Labor P.O. Box 1928 401 Broadway N.E. Albuquerque, NM 87103-1928	505-841-8408	505-841-8491	www3.state.nm.us/dol

STATE	ADDRESS	TELEPHONE	FAX	WEBSITE
NEW YORK	Department of Labor State Campus, Building 12 Albany, NY 12240	518-457-2741	518-457-6908	www.labor.state.ny.us
NORTH CAROLINA	Department of Labor 4 West Edenton Street Raleigh, NC 27601-1092	919-733-0360	919-733-6197	www.dol.state.nc.us
NORTH DAKOTA	Department of Labor State Capitol Building, 600 East Boulevard, Dept. 406 Bismark, ND 58505-0340	701-328-2660	701-328-2031	www.state.nd.us/labor
OHIO	Department of Commerce 77 South High St., 23rd Floor Columbus, OH 43266	614-466-7053	614-466-5650	http://www.state.oh.us/ohio/agency.htm
OKLAHOMA	Department of Labor 4001 N. Lincoln Blvd. Oklahoma City, OK 73105-5212	405-528-1500	405-528-5751	www.state.ok.us/~okdol
OREGON	Bureau of Labor and Industries 800 NE Oregon Street #32 Portland, OR 97232	503-731-4070	503-731-4103	www.boli.state.or.us
PENNSYLVANIA	Department of Labor and Industry 1700 Labor and Industry Building, 7th and Forster Streets Harrisburg, PA 17120	717-787-3756	717-787-8826	www.li.state.pa.us
RHODE ISLAND	Department of Labor and Training 1511 Pontiac Avenue Cranston, RI 02920	401-462-8870	401-462-8872	www.det.state.ri.us

STATE	ADDRESS	TELEPHONE	FAX	WEBSITE
SOUTH CAROLINA	Department of Labor Synergy Center - King St. Building 110 Center View Drive P.O.Box 11329 Columbia, SC 29211-1329	803-896-4300	803-896-4393	www.llr.state.sc.us
SOUTH DAKOTA	Department of Labor 700 Governors Drive Pierre, SD 57501-2291	605-773-3101	605-773-4211	www.state.sd.us/dol/dol.htm
TENNESSEE	Department of Labor Andrew Johnson Tower 710 James Robertson Pky., 8th Floor Nashville, TN 37243-0655	615-741-6642	615-741-5078	www.state.tn.us
TEXAS	Workforce Commission 101 East 15th Street, Rm 618 Austin, TX 78778	512-463-0735	512-475-2321	www.twc.state.tx.us
UTAH	Labor Commission P.O. Box 146600 Salt Lake City, UT 84111-2316	801-530-6880	801-530-6390	www.labor.state.ut.us
VERMONT	Department of Labor & Industry National Life Building, Drawer #20 Montpelier, VT 05620-3401	802-828-2288	802-828-0408	www.state.vt.us/labind
VIRGINIA	Dept. of Labor and Industry Powers-Taylor Building, 13 S. 13th Richmond, VA 23219	804-786-2377	804-371-6524	www.dli.state.va.us

STATE	ADDRESS	TELEPHONE	FAX	WEBSITE
WASHINGTON	Department of Labor & Industries P.O. Box 44001 Olympia, WA 98504-4001	360-902-4213	360-902-4202	www.wa.gov/lni
WYOMING	Department of Employment Herschler Building, 2-East 122 W. 25th Street Cheyenne, WY 82002	307-777-7672	307-777-5805	http://wydoe.state.wy.us/labstd

APPENDIX 5:
EMPLOYEE NDR FILE REQUEST FORM

Request for National Driver Register File Check on Current or Prospective Employee

Current or Prospective Employer to Receive the NDR Search Results: " **Driver Employer** " **Railroad Company**	
Employer or Agency Name	
To the specific attention of:	Business Telephone Area Code Number ( )
Mailing Address (Number and Street)	
City, State and Zip Code	

Type or Print Plainly (Avoid delays. Inquiries that cannot be read will not be processed.)

Driver's Full Legal Name (First, Middle, and Last)				
Other Names Used (Maiden, Prior Name, Nickname, Professional Name, Other)				
Mailing Address (Number and Street with Apartment Number if any or Rural Route/Carrier and Box Number)			*Home Telephone (Optional)* Area Code Number ( )	
City, State and Zip Code			*Work Telephone (Optional)* Area Code Number ( )	
Driver License Number and State (Driver must be licensed in the state initiating the search)			*Social Security Number (Optional)*	
Month, Day, and Year of Birth	Sex	Color of Eyes	Height	Weight

EMPLOYEE UNDERSTANDING: I understand that the National Driver Register (NDR) search will result in a printed report which will be sent only to the employer or regulatory agency listed above on this form. The report will indicate either (1) that the NDR does **not** contain a record matching my identification or (2) that the NDR has a probable identification (match) from one state (or more) which will be named on the report. A separate check of state files would be required (1) to verify the identification or (2) to obtain the driving record. It is the responsibility of the listed employer to obtain the state driver records and to determine or verify records which apply to me. Under the Privacy Act, I have the right to request record(s) pertaining to me from the NDR. I also understand that if convictions, suspensions or revocations of mine are found which I have not shown on my applications or interviews, I might not be hired as a driver or could lose my job as a driver, and the State where I am licensed may also take action on my driver license including suspension, cancellation, or revocation. I hereby, with my signature, authorize a one-time file search of the NDR and any resulting reports to be sent to the employer or agency named on this form.

Driver's Signature (Please read information on back before signing.)	Date

Official Use Only

Date Received	Date Sent	Internal Control

TYPE OF IDENTIFICATION: " Valid Photo Driver License " State"issued Photo ID
" Birth Certificate " Valid Passport " Valid Military ID
" Military Discharge Papers
" Other (specify) ____________________

Employee Verifying Applicant Identification (Print Name) Signature

NOTARIZATION
Required only if the NDR File Check Request is not made in person **by the current or prospective operator.**

Sworn to and ascribed before me — Notary Public Seal or Stamp

this _______ day of __________

19_________ in the city/county of

____

State of __________________

Form NDR-EMP — See Reverse Side for Additional Explanations

Requests for National Driver Register (NDR) Record Checks

Who May Obtain an NDR Record Check

Any person may ask to know whether there is an NDR record on him or her and may obtain a copy of the record if one exists. Requests from individuals require Form NDR-PRV.

Employers of drivers and locomotive engineers may also obtain NDR record checks. ***Every driver or operator on whom an NDR file check is requested is entitled to review the NDR report(s) provided to the employer.*** The results of the NDR check will be mailed only to the current or prospective employer. If no employer is named on the form or it is changed, the request will not be processed.

The following authorization applies to Railroad Company Requests

NDR CHECK AUTHORIZATION: The U. S. Department of Transportation, Federal Railroad Administration, in accordance with 49 CFR, Part 240.111, requires that I hereby request and authorize the National Highway Traffic Safety Administration (NHTSA) to perform an NDR check of my driving record for a 36-month period prior to the date of this request **including license withdrawal actions open at time of file check.** I hereby authorize the NDR to furnish a copy of the results of this NDR check directly to the railroad company identified on this inquiry form.

What NDR Records Contain

NDR results for employers will contain only the identification of the state(s) which have reported information on the driver to the NDR and only information reported within the past 3 years from the date of the inquiry. Driver control actions initiated prior to that time, even if still in effect, will not be included.

Detailed information to confirm identity or to describe the contents of the driver record can be obtained only from the State(s) listed when probable matches are reported. The name and address of the driver licensing official will be provided for each state listed.

How to Request an NDR Record Check

Using this form, which may be completed by either the current or prospective employer or the current or prospective employee, (1) the driver must authorize the request by his or her signature or mark as witnessed and (2) the driver must certify his or her identity. Any mailed NDR record check request must be notarized to certify identity.

Requests must be made to the state in which the driver is licensed.

APPENDIX 6:
DIRECTORY OF STATE NATIONAL DRIVER REGISTRY CONTACTS

STATE	ADDRESS	TELEPHONE
ALABAMA	Department of Public Safety Driver License Records Unit P.O. Box 1471 Montgomery, AL 36192-2301	(334) 242-4142
ALASKA	Department of Administration Division of Motor Vehicles Juneau Driver Licensing 2760B Sherwood Lane Juneau, AK 99801-8545	(907) 465-4363
ARIZONA	Motor Vehicle Division P.O. Box 2100 Driver Improvement Phoenix, AZ 85001	(602) 255-0072
ARKANSAS	Office of Driver Services Driver Control Section P.O. Box 1272 Little Rock, AR 72203	(501) 682-1400
CALIFORNIA	Department of Motor Vehicles Driver License Operations P.O. Box 942890 Sacramento, CA 94290-0001	(916) 657-6525
COLORADO	Motor Vehicle Division Driver Control Section 140 West 6th Avenue Denver, CO 80204	(303) 205-5613
CONNECTICUT	Department of Motor Vehicles Driver Services Division 60 State Street Wethersfield, CT 06109	(860) 263-5720

STATE	ADDRESS	TELEPHONE
DELAWARE	Division of Motor Vehicles Driver Improvement Section P.O. Box 698 Dover, DE 19903	(302) 739-4497
DISTRICT OF COLUMBIA	Bureau of Motor Vehicle Services Traffic Records and Rehab Branch 301 C Street NW., Washington, DC 20001	(202) 727-6761
FLORIDA	Bureau of Records P.O. Box 5775 Tallahassee, FL 32314-5775	(850) 487-4303
GEORGIA	Department of Public Safety Revocation Section P.O. Box 1456 Atlanta, GA 30371	(404) 657-9300
HAWAII	Department of Transportation 869 Punchbowl Street, Room 506 Honolulu, HI 96813	(808) 587-6361
IDAHO	Motor Vehicle Bureau Driver Services Section P.O. Box 7129 Boise, ID 83707	(208) 334-8736
ILLINOIS	Department of Motor Vehicles Driver Services Division 2701 South Dirksen Parkway Springfield, IL 62723	(217) 785-3108
INDIANA	Bureau of Motor Vehicles Safety Responsibility & Driver Improvement State Office Building, Room 410 Indianapolis, IN 46204	(317) 232-2840
IOWA	Driver Records 100 Euclid, Park Fair Mall P.O. Box 9204 Des Moines, IA 50306-9204	(515) 237-3086
KANSAS	Division of Vehicles Driver Control & Licensing Bureau Robert Docking State Office Building Topeka, KS 66626	(785) 296-3671
KENTUCKY	Department of Vehicle Regulation Division of Driver Licensing State Office Building, 2nd Floor Frankfort, KY 40622	(502) 564-6800

STATE	ADDRESS	TELEPHONE
LOUISIANA	Department of Public Safety Office of Motor Vehicles P.O. Box 64886 Baton Rouge, LA 70896	(225) 925-6009
MAINE	Secretary of State Motor Vehicle Division State House Station 29 Augusta, ME 04333	(207) 624-9000
MARYLAND	Motor Vehicle Administration Division of Driver Records 6601 Ritchie Highway, NE. Glen Burnie, MD 21062	(410) 768-7659
MASSACHUSETTS	Registry of Motor Vehicles, Attn: Suspensions 1135 Tremont Street Boston, MA 02120	(617) 351-7200
MICHIGAN	Department of State Bureau of Driver and Vehicle Records 7064 Crowner Drive Lansing, MI 48918	(517) 322-1460
MINNESOTA	Department of Public Safety 445 Minnesota Street St. Paul, MN 55101-5180	(651) 296-2025
MISSISSIPPI	Department of Public Safety Bureau of Driver Services P.O. Box 958 Jackson, MS 39205	(601) 987-1203
MISSOURI	Drivers License Bureau P.O. Box 200 Jefferson City, MO 65105	(573) 751-2730
MONTANA	Motor Vehicle Division Driver Services Bureau Driver Licensing Records Section 303 N. Roberts Street Helena, MT 59620	(406) 444-4590
NEBRASKA	Department of Motor Vehicles Driver Records Section P.O. Box 94789 Lincoln, NE 68509	(402) 471-3985
NEVADA	Department of Motor Vehicles Records Services Section 555 Wright Way Carson City, NV 89711-0300	(775) 684-4368

STATE	ADDRESS	TELEPHONE
NEW HAMPSHIRE	Division of Motor Vehicles Records Section James H. Hayes Safety Building Hazen Drive, Concord, NH 03305	(603) 271-3109
NEW JERSEY	Division of Motor Vehicles Driver Record Abstract Section P.O. Box 142 225 East State Street Trenton, NJ 08666	(609) 292-6500
NEW MEXICO	Division of Motor Vehicles Driver Services Bureau P.O. Box 1028 Santa Fe, NM 87504-1028	(888) 683-4636
NEW YORK	Department of Motor Vehicles Driver Licensing Division Swan Street Building Room 221 Empire State Plaza Albany, NY 12228	(518) 474-0735
NORTH CAROLINA	Division of Motor Vehicles Driver License Division 1100 New Bern Avenue Raleigh, NC 27697	(919) 715-7000
NORTH DAKOTA	State Highway Department Driver License & Traffic Safety Division 600 E. Boulevard Avenue Bismarck, ND 58505	(701) 328-2603
OHIO	Bureau of Motor Vehicles Driver License Division P.O. Box 16520 Columbus, OH 43266-0020	(614) 752-7500
OKLAHOMA	Department of Public Safety Driver Improvement Bureau P.O. Box 11415 Oklahoma City, OK 73136	(405) 425-2098
OREGON	Motor Vehicles Division Driver Licensing Section 1905 Lana Avenue NE. Salem, OR 97314	(503) 945-5400
PENNSYLVANIA	Bureau of Driver Licensing Information Sales Unit P.O. Box 8691 Harrisburg, PA 17105	(717) 787-7154

STATE	ADDRESS	TELEPHONE
RHODE ISLAND	Registry of Motor Vehicles Operator Control Section 286 Main Street Pawtucket, RI 02860	(401) 721-2650
SOUTH CAROLINA	South Carolina Department of Public Safety Driver Records P.O. Box 100178 Columbia, SC 29202-3178	(800) 442-1368
SOUTH DAKOTA	Department of Commerce & Regulation Driver Improvement Program 118 W. Capitol Avenue Peirre, SD 57501-2036	(605) 773-6883
TENNESSEE	Department of Safety Driver Control Division 1150 Foster Avenue Nashville, TN 37210	(615) 251-5166
TEXAS	Department of Public Safety Driver Improvement and Control P.O. Box 4087 Austin, TX 78773	(512) 424-2600
UTAH	Motor Vehicle Division Motor Vehicle Records Department P.O. Box 30560 Salt Lake City UT 84130-0560	(801) 965-3872
VERMONT	Department of Motor Vehicles Driver Improvement 120 State Street Montpelier, VT 05603	(802) 828-2155
VIRGINIA	Department of Motor Vehicles Driver Licensing & Info Division P.O. Box 27412 Richmond, VA 23269	(804) 367-0538
WASHINGTON	Department of Licensing Division of Driver Services P.O. Box 9030 Olympia, WA 98504	(360) 902-3900
WEST VIRGINIA	Department of Motor Vehicles Driver Improvement Division 1800 Washington Street East Charleston, WV 25317	(304) 558-2413

Source: National Highway Transportation Safety Administration.

APPENDIX 7: EMPLOYMENT ELIGIBILITY VERIFICATION FORM (FORM I-9)

U.S. Department of Justice
Immigration and Naturalization Service

OMB No. 1115-0136
Employment Eligibility Verification

INSTRUCTIONS

PLEASE READ ALL INSTRUCTIONS CAREFULLY BEFORE COMPLETING THIS FORM.

Anti-Discrimination Notice. It is illegal to discriminate against any individual (other than an alien not authorized to work in the U.S.) in hiring, discharging, or recruiting or referring for a fee because of that individual's national origin or citizenship status. It is illegal to discriminate against work eligible individuals. Employers **CANNOT** specify which document(s) they will accept from an employee. The refusal to hire an individual because of a future expiration date may also constitute illegal discrimination.

Section 1 - Employee. All employees, citizens and noncitizens, hired after November 6, 1986, must complete Section 1 of this form at the time of hire, which is the actual beginning of employment. **The employer is responsible for ensuring that Section 1 is timely and properly completed.**

Preparer/Translator Certification. The Preparer/Translator Certification must be completed if Section 1 is prepared by a person other than the employee. A preparer/translator may be used only when the employee is unable to complete Section 1 on his/her own. However, the employee must still sign Section 1.

Section 2 - Employer. For the purpose of completing this form, the term "employer" includes those recruiters and referrers for a fee who are agricultural associations, agricultural employers or farm labor contractors.

Employers must complete Section 2 by examining evidence of identity and employment eligibility within three (3) business days of the date employment begins. If employees are authorized to work, but are unable to present the required document(s) within three business days, they must present a receipt for the application of the document(s) within three business days and the actual document(s) within ninety (90) days. However, if employers hire individuals for a duration of less than three business days, Section 2 must be completed at the time employment begins. **Employers must record: 1)** document title; **2)** issuing authority; **3)** document number, **4)** expiration date, if any; and **5)** the date employment begins. Employers must sign and date the certification. Employees must present original documents. Employers may, but are not required to, photocopy the document(s) presented. These photocopies may only be used for the verification process and must be retained with the I-9. **However, employers are still responsible for completing the I-9.**

Section 3 - Updating and Reverification. Employers must complete Section 3 when updating and/or reverifying the I-9. Employers must reverify employment eligibility of their employees on or before the expiration date recorded in Section 1. Employers **CANNOT** specify which document(s) they will accept from an employee.

- If an employee's name has changed at the time this form is being updated/ reverified, complete Block A.
- If an employee is rehired within three (3) years of the date this form was originally completed and the employee is still eligible to be employed on the same basis as previously indicated on this form (updating), complete Block B and the signature block.
- If an employee is rehired within three (3) years of the date this form was originally completed and the employee's work authorization has expired **or** if a current employee's work authorization is about to expire (reverification), complete Block B and:
  - examine any document that reflects that the employee is authorized to work in the U.S. (see List A **or** C).
  - record the document title, document number and expiration date (if any) in Block C, and
  - complete the signature block.

Photocopying and Retaining Form I-9. A blank I-9 may be reproduced, provided both sides are copied. The Instructions must be available to all employees completing this form. Employers must retain completed I-9s for three (3) years after the date of hire or one (1) year after the date employment ends, whichever is later.

For more detailed information, you may refer to the INS Handbook for Employers, (Form M-274). You may obtain the handbook at your local INS office.

Privacy Act Notice. The authority for collecting this information is the Immigration Reform and Control Act of 1986, Pub. L. 99-603 (8 USC 1324a).

This information is for employers to verify the eligibility of individuals for employment to preclude the unlawful hiring, or recruiting or referring for a fee, of aliens who are not authorized to work in the United States.

This information will be used by employers as a record of their basis for determining eligibility of an employee to work in the United States. The form will be kept by the employer and made available for inspection by officials of the U.S. Immigration and Naturalization Service, the Department of Labor and the Office of Special Counsel for Immigration Related Unfair Employment Practices.

Submission of the information required in this form is voluntary. However, an individual may not begin employment unless this form is completed, since employers are subject to civil or criminal penalties if they do not comply with the Immigration Reform and Control Act of 1986.

Reporting Burden. We try to create forms and instructions that are accurate, can be easily understood and which impose the least possible burden on you to provide us with information. Often this is difficult because some immigration laws are very complex. Accordingly, the reporting burden for this collection of information is computed as follows: **1)** learning about this form, 5 minutes; **2)** completing the form, 5 minutes; and **3)** assembling and filing (recordkeeping) the form, 5 minutes, for an average of 15 minutes per response. If you have comments regarding the accuracy of this burden estimate, or suggestions for making this form simpler, you can write to the Immigration and Naturalization Service, HQPDI, 425 I Street, N.W., Room 4034, Washington, DC 20536. OMB No. 1115-0136.

EMPLOYERS MUST RETAIN COMPLETED FORM I-9
PLEASE DO NOT MAIL COMPLETED FORM I-9 TO INS

Form I-9 (Rev. 11-21-91)N

U.S. Department of Justice
Immigration and Naturalization Service

OMB No. 1115-0136

Employment Eligibility Verification

Please read instructions carefully before completing this form. The instructions must be available during completion of this form. ANTI-DISCRIMINATION NOTICE: It is illegal to discriminate against work eligible individuals. Employers CANNOT specify which document(s) they will accept from an employee. The refusal to hire an individual because of a future expiration date may also constitute illegal discrimination.

Section 1. Employee Information and Verification. To be completed and signed by employee at the time employment begins.

Print Name: Last	First	Middle Initial	Maiden Name
Address *(Street Name and Number)*		Apt. #	Date of Birth *(month/day/year)*
City	State	Zip Code	Social Security #

I am aware that federal law provides for imprisonment and/or fines for false statements or use of false documents in connection with the completion of this form.

I attest, under penalty of perjury, that I am (check one of the following):
- ☐ A citizen or national of the United States
- ☐ A Lawful Permanent Resident (Alien # A__________)
- ☐ An alien authorized to work until ___/___/___
  (Alien # or Admission #) ______________

Employee's Signature	Date *(month/day/year)*

Preparer and/or Translator Certification. *(To be completed and signed if Section 1 is prepared by a person other than the employee.) I attest, under penalty of perjury, that I have assisted in the completion of this form and that to the best of my knowledge the information is true and correct.*

Preparer's/Translator's Signature	Print Name
Address *(Street Name and Number, City, State, Zip Code)*	Date *(month/day/year)*

Section 2. Employer Review and Verification. To be completed and signed by employer. Examine one document from List A OR examine one document from List B and one from List C, as listed on the reverse of this form, and record the title, number and expiration date, if any, of the document(s)

List A	OR	List B	AND	List C
Document title: __________		__________		__________
Issuing authority: __________		__________		__________
Document #: __________		__________		__________
Expiration Date *(if any)*: ___/___/___		___/___/___		___/___/___
Document #: __________				
Expiration Date *(if any)*: ___/___/___				

CERTIFICATION - I attest, under penalty of perjury, that I have examined the document(s) presented by the above-named employee, that the above-listed document(s) appear to be genuine and to relate to the employee named, that the employee began employment on *(month/day/year)* ___/___/___ **and that to the best of my knowledge the employee is eligible to work in the United States. (State employment agencies may omit the date the employee began employment.)**

Signature of Employer or Authorized Representative	Print Name	Title
Business or Organization Name	Address *(Street Name and Number, City, State, Zip Code)*	Date *(month/day/year)*

Section 3. Updating and Reverification. To be completed and signed by employer.

A. New Name *(if applicable)*	B. Date of rehire *(month/day/year) (if applicable)*

C. If employee's previous grant of work authorization has expired, provide the information below for the document that establishes current employment eligibility.

Document Title: __________ Document #: __________ Expiration Date (if any): ___/___/___

I attest, under penalty of perjury, that to the best of my knowledge, this employee is eligible to work in the United States, and if the employee presented document(s), the document(s) I have examined appear to be genuine and to relate to the individual.

Signature of Employer or Authorized Representative	Date *(month/day/year)*

Form I-9 (Rev. 11-21-91)N Page 2

LISTS OF ACCEPTABLE DOCUMENTS

LIST A

Documents that Establish Both Identity and Employment Eligibility

1. U.S. Passport (unexpired or expired)
2. Certificate of U.S. Citizenship *(INS Form N-560 or N-561)*
3. Certificate of Naturalization *(INS Form N-550 or N-570)*
4. Unexpired foreign passport, with *I-551 stamp or* attached *INS Form I-94* indicating unexpired employment authorization
5. Permanent Resident Card or Alien Registration Receipt Card with photograph *(INS Form I-151 or I-551)*
6. Unexpired Temporary Resident Card *(INS Form I-688)*
7. Unexpired Employment Authorization Card *(INS Form I-688A)*
8. Unexpired Reentry Permit *(INS Form I-327)*
9. Unexpired Refugee Travel Document *(INS Form I-571)*
10. Unexpired Employment Authorization Document issued by the INS which contains a photograph *(INS Form I-688B)*

OR

LIST B

Documents that Establish Identity

1. Driver's license or ID card issued by a state or outlying possession of the United States provided it contains a photograph or information such as name, date of birth, gender, height, eye color and address
2. ID card issued by federal, state or local government agencies or entities, provided it contains a photograph or information such as name, date of birth, gender, height, eye color and address
3. School ID card with a photograph
4. Voter's registration card
5. U.S. Military card or draft record
6. Military dependent's ID card
7. U.S. Coast Guard Merchant Mariner Card
8. Native American tribal document
9. Driver's license issued by a Canadian government authority

For persons under age 18 who are unable to present a document listed above:

10. School record or report card
11. Clinic, doctor or hospital record
12. Day-care or nursery school record

AND

LIST C

Documents that Establish Employment Eligibility

1. U.S. social security card issued by the Social Security Administration *(other than a card stating it is not valid for employment)*
2. Certification of Birth Abroad issued by the Department of State *(Form FS-545 or Form DS-1350)*
3. Original or certified copy of a birth certificate issued by a state, county, municipal authority or outlying possession of the United States bearing an official seal
4. Native American tribal document
5. U.S. Citizen ID Card *(INS Form I-197)*
6. ID Card for use of Resident Citizen in the United States *(INS Form I-179)*
7. Unexpired employment authorization document issued by the INS *(other than those listed under List A)*

Illustrations of many of these documents appear in Part 8 of the Handbook for Employers (M-274)

Form I-9 (Rev. 10/4/00)Y Page 3

APPENDIX 8: DIRECTORY OF STATE WORKERS' COMPENSATION BOARDS

STATE	ADDRESS	TELEPHONE
ALABAMA	Workers' Compensation Division Department of Industrial Relations Industrial Relations Building Montgomery AL 36131	(334) 242-2868
ALASKA	Workers' Compensation Division Department of Labor 1111 West 8th Street, Suite 307 P.O. Box 25512 Juneau AK 99802-5512	(907) 465-2790
ARIZONA	Industrial Commission 800 West Washington Street P. O. Box 19070 Phoenix AZ 85005-9070	(602) 542-4411
ARKANSAS	Workers' Compensation Commission 4th and Spring Streets P. O. Box 950 Little Rock, AR 72203-9050	(501) 682-3930
CALIFORNIA	Division of Workers' Compensation 45 Fremont Street, Suite 3160 P. O. Box 420603 San Francisco, CA 94142	(415) 975-0700
COLORADO	Division of Workers' Compensation 1515 Arapahoe Street Denver, CO 80202	(303) 575-8700
CONNECTICUT	Workers' Compensation Commission 21 Oak Street Hartford, CT 06106	(860) 493-1500
DELAWARE	Industrial Accident Board 4425 N. Market Street, 3rd Floor Wilmington, DE 19802	(302) 761-8200

DISTRICT OF COLUMBIA	Office of Workers' Compensation 1200 Upshur St. NW, 2nd Floor Washington, DC 20011	(202) 576-6265
FLORIDA	Division of Workers' Compensation 301 Forrest Building 2728 Centerview Drive Tallahassee, FL 32399-0680	(904) 488-2514
GEORGIA	Board of Workers' Compensation 270 Peachtree Street NW Atlanta, GA 30303-1205	(404) 656-3875
HAWAII	Disability Compensation Division Department of Labor and Industrial Relations 830 Punchbowl Street, Room 211 P. O. Box 3769 Honolulu, HI 96812	(808) 586-9151
IDAHO	Industrial Commission Statehouse Mail 317 Main Street P. O. Box 83720 Boise, ID 83720-0041	(208) 334-6000
ILLINOIS	Industrial Commission 100 West Randolph Street, Suite 8-200 Chicago, IL 60601	(312) 814-6500
INDIANA	Workers' Compensation Board Room W-196 402 West Washington Street Indianapolis, IN 46204	(317) 232-3808
IOWA	Division of Industrial Services Iowa Workforce Development 1000 East Grand Avenue Des Moines, IA 50319	(515) 281-5934
KANSAS	Division of Workers' Compensation Department of Human Resources 800 SW Jackson Street, Suite 600 Topeka, KS 66612-1227	(913) 296-3441
KENTUCKY	Department of Workers Claims 1270 Louisville Road Perimeter Park West, Bldg. C Frankfort, KY 40601	(502) 564-5550
LOUISIANA	Office of Workers' Compensation 1001 North 23rd Street P. O. Box 94040 Baton Rouge, LA 70802-9040	(504) 342-7555
MAINE	Workers' Compensation Board 27 State House Station Augusta, ME 04333	(207) 287-3751

MARYLAND	Workers' Compensation Commission 6 North Liberty Street Baltimore, MD 21201-3785	(410) 767-0900
MASSACHUSETTS	Dept. of Industrial Accidents 600 Washington Street, 7th Floor Boston, MA 02111	(617) 727-4900
MICHIGAN	Bureau of Workers' Disability Compensation 201 North Washington Square P. O. Box 30016 Lansing, MI 48909	(517) 322-1296
MINNESOTA	Division of Workers' Compensation Department of Labor and Industry 443 Lafayette Road North St. Paul, MN 55155-4319	(612) 296-6490
MISSISSIPPI	Workers' Compensation Commission 1428 Lakeland Drive P. O. Box 5300 Jackson, MS 39296-5300	(601) 987-4200
MISSOURI	Division of Workers' Compensation Department of Labor and Industrial Relations P. O. Box 58, Jefferson City, MO 65102	(573) 751-4231
MONTANA	Employment Relations Division Department of Labor and Industry 1805 Prospect Avenue P. O. Box 8011 Helena, MT 59604-8011	(406) 444-6530
NEBRASKA	Workers' Compensation Court Capitol Building P. O. Box 98908 Lincoln, NE 68509-8908	(402) 471-2568
NEVADA	State Industrial Insurance System 515 East Musser Street Carson City, NV 89714	(702) 687-5220
NEW HAMPSHIRE	Workers' Compensation Division Department of Labor 95 Pleasant Street Concord, NH 03301	(603) 271-3176
NEW JERSEY	Division of Workers' Compensation Department of Labor, C. N. 381 Trenton, NJ 08625-0381	(609) 292-2414
NEW MEXICO	Workers' Compensation Admin. 2410 Centre Drive SE P. O. Box 27198 Albuquerque, NM 87125-7198	(505) 841-6000

NEW YORK	Workers' Compensation Board 100 Broadway-Menands Albany, NY 12241	(518) 474-6670
NORTH CAROLINA	Industrial Commission Dobbs Building 430 North Salisbury Street Raleigh, NC 27611	(919) 733-4820
NORTH DAKOTA	Workers' Compensation Bureau 500 East Front Avenue Bismarck, ND 58504-5685	(701) 328-3800
OHIO	Bureau of Workers' Compensation 30 West Spring Street Columbus, OH 43266-0581	(614) 466-8751
OKLAHOMA	Workers' Compensation Court 1915 North Stiles Avenue Oklahoma City, OK 73105-4904	(405) 557-7600
OREGON	Workers' Compensation Division 350 Winter Street NE Salem, OR 97310	(503) 945-7881
PENNSYLVANIA	Bureau of Workers' Compensation Department of Labor and Industry 1171 So. Cameron Street, Rm. 103 Harrisburg, PA 17104-2501	(717) 783-5421
PUERTO RICO	Industrial Commission G.P.O. Box 364466 San Juan, PR 00936-4466	(809) 783-2028
RHODE ISLAND	Division of Workers' Compensation Department of Labor 610 Manton Avenue P.O. Box 3500 Providence, RI 02909	(401) 457-1800
SOUTH CAROLINA	Workers' Compensation Commission 1612 Marion Street P. O. Box 1715 Columbia, SC 29202-1715	(803) 737-5700
SOUTH DAKOTA	Division of Labor & Management Department of Labor 700 Governors Dr., Kneip Bldg. Pierre, SD 57501-2291	(605) 773-3681
TENNESSEE	Division of Workers' Compensation Department of Labor Gateway Plaza, 2nd Floor 710 James Robertson Parkway Nashville,. TN 37243-0661	(615) 741-2395

TEXAS	Workers' Compensation Commission Southfield Building 4000 South IH-35 Austin, TX 78704-7491	(512) 448-7900
UTAH	Industrial Accident Division Industrial Commission 160 East 300 South, 3rd Floor P. O. Box 146610 Salt Lake City, UT 84114-6610	(801) 530-6800
VERMONT	Department of Labor and Industry National Life Building Drawer 20 Montpelier, VT 05620-3401	(802) 828-2286
VIRGINIA	Workers' Compensation Commission 1000 DMV Drive Richmond, VA 23220	(804) 367-8600
VIRGIN ISLANDS	Workers' Compensation Division Department of Labor 2131 Hospital Street St. Croix, VI 00820-4666	(809) 692-9390
WASHINGTON	Department of Labor and Industries 7273 Linderson Way, SW Tumwater, WA 98501-4850	(360) 902-5800
WEST VIRGINIA	Workers' Compensation Division 4700 MacCorkle Avenue SE Charleston, WV 25304	(304) 926-5048
WISCONSIN	Workers' Compensation Division Department of Workforce Development 201 East Washington Avenue P. O. Box 7901 Madison, WI 53707	(608) 266-1340
WYOMING	Workers' Compensation Division 122 West 25th Street, 2nd Floor Herschler Building, East Wing Cheyenne, WY 82002-0700	(307) 777-7159

APPENDIX 9:
DIRECTORY OF STATE UNEMPLOYMENT TAX AGENCIES

STATE	ADDRESS	TELEPHONE	WEBSITE
ALABAMA	Department of Industrial Relations 649 Monroe Street Montgomery, AL 36131	334-242-8467	www.dir.state.al.us
ALASKA	Department of Labor Employment Security Division P.O. Box 25506 Juneau, AK 99802-5509	888-448-3527	www.labor.state.ak.us
ARIZONA	Department of Economic Security 3225 North Central Avenue, Suite 1411 Phoenix, AZ 85012	602-248-9396	www.arizona.gov
ARKANSAS	Employment Security Department P.O. Box 2981 Little Rock, AR 72203	501-682-3253	www.arkansas.gov

STATE	ADDRESS	TELEPHONE	WEBSITE
CALIFORNIA	Employment Development Department 3321 Power Inn Road, Suite 220 Sacramento, CA 95826-6110	877-547-4503	www.edd.cahwnet.gov
COLORADO	Department of Labor and Employment, 1515 Arapahoe Street, Suite 400 Denver, CO 80202-2117	800-480-8299	www.coworkforce.com
CONNECTICUT	Labor Department Employment Security Division 200 Folly Brook Boulevard Wethersfield, CT 06109-1114	860-263-6550	www.ct.gov
DELAWARE	Department of Labor 4425 N. Market Street Wilmington, DE 19802	302-761-8484	www.delawareworks.com
DISTRICT OF COLUMBIA	Department of Employment Services 609 H Street NE, Room 362 Washington, D.C. 20001	202-698-7550	www.does.dc.gov
FLORIDA	Agency for Workforce Innovation 107 E. Madison Street, MSC 229 Tallahassee, FL 32399-0100	800-352-3671	www.floridajobs.org
GEORGIA	Department of Labor 148 International Blvd., Suite 800 Atlanta, GA 30303	404-232-3301	www.dol.state.ga.us
HAWAII	Department of Labor and Industrial Relations 830 Punchbowl Street, Room 437 Honolulu, HI 96813	808-586-8913	www.dlir.state.hi.us

STATE	ADDRESS	TELEPHONE	WEBSITE
IDAHO	Department of Labor 317 W. Main Street Boise, ID 83735	208-332-3576	www.labor.state.id.us
ILLINOIS	Department of Employment Security 33 South State Street Chicago, IL 60603	312-793-1900	www.ides.state.il.us
INDIANA	Department of Workforce Development 10 North Senate Avenue Indianapolis, IN 46204	317-232-7436	www.in.gov/dwd/unemploy
IOWA	Workforce Development 1000 East Grand Avenue Des Moines, IA 50319	515-281-5339	www.iowaworkforce.org/ui
KANSAS	Department of Human Resources 401 S.W. Topeka Boulevard Topeka, KS 66603	785-296-5025	www.dol.ks.gov
KENTUCKY	Division of Employment Services P.O. Box 948 Frankfort, KY 40602	502-564-6838	www.oet.ky.gov
LOUISIANA	Department of Labor P.O. Box 98146 Baton Rouge, LA 70804	225-342-2944	www.ldol.state.la.us
MAINE	Department of Labor P.O. Box 259 Augusta, ME 04332-0259	207-287-3176	www.maine.gov

STATE	ADDRESS	TELEPHONE	WEBSITE
MARYLAND	Department of Labor Licensing and Regulation 1100 Eutaw St., Room 414 Baltimore, MD 21201	800-492-5524	www.maryland.gov
MASSACHUSETTS	Division of Employment and Training 19 Staniford Street Boston, MA 02114	617-626-5050	www.detma.org
MICHIGAN	Department of Labor & Economic Growtn Unemployment Insurance Agency 3024 West Grand Blvd. Detroit, MI 48202	313-456-2180	www.mikchigan.gov/uia
MINNESOTA	Department of Labor and Economic Security 390 North Robert Street St. Paul, MN 55101	651-296-6141	www.uimn.org
MISSISSIPPI	Employment Security Commission P.O. Box 22781 Jackson, MS 39215-1699	601-961-7755	www.mesc.state.ms.us/tax
MISSOURI	Division of Employment Security P.O. Box 59 Jefferson City, MO 65104	573-751-3340	www.missouri.gov
MONTANA	Department of Labor and Industry P.O. Box 1728 Helena, MT 59624-1728	406-444-6900	www.dli.state.mt.us

STATE	ADDRESS	TELEPHONE	WEBSITE
NEBRASKA	Department of Labor P.O. Box 94600 State House Station Lincoln, NE 68509	402-471-9899	www.dol.state.ne.us
NEVADA	Department of Employment Training and Rehabiliation 500 East Third Street Carson City, NV 89713	775-687-4545	www.detr.state.nv.us
NEW HAMPSHIRE	Department of Employment Security 32 South Main Street Concord, NH 03301	603-228-4033	www.nhes.state.nh.us
NEW JERSEY	Department of Labor P.O. Box 947 Trenton, NJ 08625-0947	609-633-6400	www.nj.gov
NEW MEXICO	Department of Labor P.O. Box 2281 Albuquerque, NM 87103	505-841-8582	www.dol.state.nm.us
NEW YORK	Department of Labor State Campus, Building 12 Albany, NY 12240	888-899-8810	www.labor.state.ny.us
NORTH CAROLINA	Employment Security Commission P.O. Box 26504 Raleigh, NC 27611	919-733-07396	www.ncesc.com
NORTH DAKOTA	Job Service of North Dakota P.O. Box 5507 Bismark, ND 58506-5507	800-472-2952	www.state.nd.us

STATE	ADDRESS	TELEPHONE	WEBSITE
OHIO	Department of Job and Family Services 4300 Kimberly Parkway Columbus, OH 43232	614-466-2319	www.state.oh.us
OKLAHOMA	Employment Security Commission 2401 North Lincoln Oklahoma City, OK 73105	405-557-7170	www.oesc.state.ok.us/ui
OREGON	Employment Department 875 Union Street NE Salem, OR 97311	503-947-1488	www.emp.state.or.us
PENNSYLVANIA	Department of Labor and Industry 7th and Forster Streets Harrisburg, PA 17121	888-313-7284	www.dli.state.pa.us
RHODE ISLAND	Division of Taxation One Capitol Hill Providence, RI 02908-5829	401-222-3696	www.det.state.ri.us
SOUTH CAROLINA	Employment Security Commission P.O. Box 995 Columbia, SC 29202	803-737-3075	www.sces.org/ui
SOUTH DAKOTA	Department of Labor P.O. Box 4730 Aberdeen, SD 57402	605-626-2312	www.state.sd.us
TENNESSEE	Department of Labor and Workforce 500 James Robertson Pky., 9th Floor Nashville, TN 37245-3500	615-741-2486	www.state.tn.us

STATE	ADDRESS	TELEPHONE	WEBSITE
TEXAS	Workforce Commission 101 East 15th Street Austin, TX 78778	512-463-2700	www.twc.state.tx.us
UTAH	Department of Workforce Services P.O. Box 143001 Salt Lake City, UT 84144	801-526-9400	www.jobs.utah.gov/ui
VERMONT	Department of Employment and Training, P.O. Box 488 Montpelier, VT 05601-0488	802-828-4000	www.det.state.vt.us
VIRGINIA	Employment Commission P.O. Box 1358 Richmond, VA 23218-1358	804-371-6325	www.vec.state.va.us
WASHINGTON	Employment Security Department P.O. Box 9046 Olympia, WA 98507-9046	360-902-9360	www.wa.gov/esd
WEST VIRGINIA	Bureau of Employment Programs 112 California Avenue Charleston, WV 25305-0112	304-558-2675	www.wv.gov
WISCONSIN	Department of Workforce Development P.O. Box 7942 Madison, WI 53707-7946	608-261-6700	www.dwd.state.wi.us
WYOMING	Department of Employment P.O. Box 2760 Casper, WY 82602	307-235-3217	wydoe.state.wy.us

APPENDIX 10
EMPLOYEE'S WITHHOLDING ALLOWANCE CERTIFICATE (IRS FORM W-4)

Form W-4 (2005)

Purpose. Complete Form W-4 so that your employer can withhold the correct federal income tax from your pay. Because your tax situation may change, you may want to refigure your withholding each year.

Exemption from withholding. If you are exempt, complete only lines 1, 2, 3, 4, and 7 and sign the form to validate it. Your exemption for 2005 expires February 16, 2006. See Pub. 505, Tax Withholding and Estimated Tax.

Note. You cannot claim exemption from withholding if (a) your income exceeds $800 and includes more than $250 of unearned income (for example, interest and dividends) and (b) another person can claim you as a dependent on their tax return.

Basic instructions. If you are not exempt, complete the **Personal Allowances Worksheet** below. The worksheets on page 2 adjust your withholding allowances based on itemized deductions, certain credits, adjustments to income, or two-earner/two-job situations. Complete all worksheets that apply. However, you may claim fewer (or zero) allowances.

Head of household. Generally, you may claim head of household filing status on your tax return only if you are unmarried and pay more than 50% of the costs of keeping up a home for yourself and your dependent(s) or other qualifying individuals. See line **E** below.

Tax credits. You can take projected tax credits into account in figuring your allowable number of withholding allowances. Credits for child or dependent care expenses and the child tax credit may be claimed using the **Personal Allowances Worksheet** below. See Pub. 919, How Do I Adjust My Tax Withholding? for information on converting your other credits into withholding allowances.

Nonwage income. If you have a large amount of nonwage income, such as interest or dividends, consider making estimated tax payments using Form 1040-ES, Estimated Tax for Individuals. Otherwise, you may owe additional tax.

Two earners/two jobs. If you have a working spouse or more than one job, figure the total number of allowances you are entitled to claim on all jobs using worksheets from only one Form W-4. Your withholding usually will be most accurate when all allowances are claimed on the Form W-4 for the highest paying job and zero allowances are claimed on the others.

Nonresident alien. If you are a nonresident alien, see the Instructions for Form 8233 before completing this Form W-4.

Check your withholding. After your Form W-4 takes effect, use Pub. 919 to see how the dollar amount you are having withheld compares to your projected total tax for 2005. See Pub. 919, especially if your earnings exceed $125,000 (Single) or $175,000 (Married).

Recent name change? If your name on line 1 differs from that shown on your social security card, call 1-800-772-1213 to initiate a name change and obtain a social security card showing your correct name.

Personal Allowances Worksheet (Keep for your records.)

A Enter "1" for **yourself** if no one else can claim you as a dependent ______ A ____

B Enter "1" if:
- You are single and have only one job; or
- You are married, have only one job, and your spouse does not work; or
- Your wages from a second job or your spouse's wages (or the total of both) are $1,000 or less.

____ B ____

C Enter "1" for your **spouse.** But, you may choose to enter "-0-" if you are married and have either a working spouse or more than one job. (Entering "-0-" may help you avoid having too little tax withheld.) ______ C ____

D Enter number of **dependents** (other than your spouse or yourself) you will claim on your tax return ______ D ____

E Enter "1" if you will file as **head of household** on your tax return (see conditions under **Head of household** above) __ E ____

F Enter "1" if you have at least $1,500 of **child or dependent care expenses** for which you plan to claim a credit ____ F ____
(**Note.** Do **not** include child support payments. See **Pub. 503,** Child and Dependent Care Expenses, for details.)

G **Child Tax Credit** (including additional child tax credit):
- If your total income will be less than $54,000 ($79,000 if married), enter "2" for each eligible child.
- If your total income will be between $54,000 and $84,000 ($79,000 and $119,000 if married), enter "1" for each eligible child plus "1" **additional** if you have four or more eligible children.

G ____

H Add lines A through G and enter total here. (**Note.** This may be different from the number of exemptions you claim on your tax return.) ▶ H ____

For accuracy, complete all worksheets that apply.
- If you plan to **itemize or claim adjustments to income** and want to reduce your withholding, see the **Deductions and Adjustments Worksheet** on page 2.
- If you have **more than one job** or are **married and you and your spouse both work** and the combined earnings from all jobs exceed $35,000 ($25,000 if married) see the **Two-Earner/Two-Job Worksheet** on page 2 to avoid having too little tax withheld.
- If **neither** of the above situations applies, **stop here** and enter the number from line H on line 5 of Form W-4 below.

Cut here and give Form W-4 to your employer. Keep the top part for your records.

Form **W-4**
Department of the Treasury
Internal Revenue Service

Employee's Withholding Allowance Certificate

▶ **Whether you are entitled to claim a certain number of allowances or exemption from withholding is subject to review by the IRS. Your employer may be required to send a copy of this form to the IRS.**

OMB No. 1545-0010
2005

1 Type or print your first name and middle initial | Last name | 2 Your social security number

Home address (number and street or rural route)

3 ☐ Single ☐ Married ☐ Married, but withhold at higher Single rate.
Note. If married, but legally separated, or spouse is a nonresident alien, check the "Single" box.

City or town, state, and ZIP code

4 **If your last name differs from that shown on your social security card, check here. You must call 1-800-772-1213 for a new card.** ▶ ☐

5 Total number of allowances you are claiming (from line **H** above **or** from the applicable worksheet on page 2) | 5

6 Additional amount, if any, you want withheld from each paycheck | 6 $

7 I claim exemption from withholding for 2005, and I certify that I meet **both** of the following conditions for exemption.
- Last year I had a right to a refund of **all** federal income tax withheld because I had **no** tax liability **and**
- This year I expect a refund of **all** federal income tax withheld because I expect to have **no** tax liability.

If you meet both conditions, write "Exempt" here ▶ | 7

Under penalties of perjury, I declare that I have examined this certificate and to the best of my knowledge and belief, it is true, correct, and complete.

Employee's signature
(Form is not valid unless you sign it.) ▶ **Date** ▶

8 Employer's name and address (Employer: Complete lines 8 and 10 only if sending to the IRS.) | 9 Office code (optional) | 10 Employer identification number (EIN)

For Privacy Act and Paperwork Reduction Act Notice, see page 2. Cat. No. 10220Q Form **W-4** (2005)

Form W-4 (2005) Page **2**

Deductions and Adjustments Worksheet

Note. Use this worksheet *only* if you plan to itemize deductions, claim certain credits, or claim adjustments to income on your 2005 tax return.

1 Enter an estimate of your 2005 itemized deductions. These include qualifying home mortgage interest, charitable contributions, state and local taxes, medical expenses in excess of 7.5% of your income, and miscellaneous deductions. (For 2005, you may have to reduce your itemized deductions if your income is over $145,950 ($72,975 if married filing separately). See *Worksheet 3* in Pub. 919 for details.) ______ **1** $ ______

2 Enter: { $10,000 if married filing jointly or qualifying widow(er); $ 7,300 if head of household; $ 5,000 if single or married filing separately } ______ **2** $ ______

3 **Subtract** line 2 from line 1. If line 2 is greater than line 1, enter "-0-" ______ **3** $ ______

4 Enter an estimate of your 2005 adjustments to income, including alimony, deductible IRA contributions, and student loan interest **4** $ ______

5 **Add** lines 3 and 4 and enter the total. (Include any amount for credits from *Worksheet 7* in Pub. 919) **5** $ ______

6 Enter an estimate of your 2005 nonwage income (such as dividends or interest) ______ **6** $ ______

7 **Subtract** line 6 from line 5. Enter the result, but not less than "-0-" ______ **7** $ ______

8 **Divide** the amount on line 7 by $3,200 and enter the result here. Drop any fraction ______ **8** ______

9 Enter the number from the **Personal Allowances Worksheet,** line H, page 1 ______ **9** ______

10 **Add** lines 8 and 9 and enter the total here. If you plan to use the **Two-Earner/Two-Job Worksheet,** also enter this total on line 1 below. Otherwise, **stop here** and enter this total on Form W-4, line 5, page 1 **10** ______

Two-Earner/Two-Job Worksheet (See *Two earners/two jobs* on page 1.)

Note. Use this worksheet *only* if the instructions under line H on page 1 direct you here.

1 Enter the number from line H, page 1 (or from line 10 above if you used the **Deductions and Adjustments Worksheet**) **1** ______

2 Find the number in **Table 1** below that applies to the **LOWEST** paying job and enter it here ______ **2** ______

3 If line 1 is **more than or equal to** line 2, subtract line 2 from line 1. Enter the result here (if zero, enter "-0-") and on Form W-4, line 5, page 1. **Do not** use the rest of this worksheet ______ **3** ______

Note. If line 1 is *less than* line 2, enter "-0-" on Form W-4, line 5, page 1. Complete lines 4–9 below to calculate the additional withholding amount necessary to avoid a year-end tax bill.

4 Enter the number from line 2 of this worksheet ______ **4** ______

5 Enter the number from line 1 of this worksheet ______ **5** ______

6 **Subtract** line 5 from line 4 ______ **6** ______

7 Find the amount in **Table 2** below that applies to the **HIGHEST** paying job and enter it here ______ **7** $ ______

8 **Multiply** line 7 by line 6 and enter the result here. This is the additional annual withholding needed ______ **8** $ ______

9 Divide line 8 by the number of pay periods remaining in 2005. For example, divide by 26 if you are paid every two weeks and you complete this form in December 2004. Enter the result here and on Form W-4, line 6, page 1. This is the additional amount to be withheld from each paycheck ______ **9** $

Table 1: Two-Earner/Two-Job Worksheet

Married Filing Jointly						All Others	
If wages from **HIGHEST** paying job are—	AND, wages from **LOWEST** paying job are—	Enter on line 2 above	If wages from **HIGHEST** paying job are—	AND, wages from **LOWEST** paying job are—	Enter on line 2 above	If wages from **LOWEST** paying job are—	Enter on line 2 above
$0 - $40,000	$0 - $4,000	0	$40,001 and over	30,001 - 36,000	6	$0 - $6,000	0
	4,001 - 8,000	1		36,001 - 45,000	7	6,001 - 12,000	1
	8,001 - 18,000	2		45,001 - 50,000	8	12,001 - 18,000	2
	18,001 and over	3		50,001 - 60,000	9	18,001 - 24,000	3
				60,001 - 65,000	10	24,001 - 31,000	4
$40,001 and over	$0 - $4,000	0		65,001 - 75,000	11	31,001 - 45,000	5
	4,001 - 8,000	1		75,001 - 90,000	12	45,001 - 60,000	6
	8,001 - 18,000	2		90,001 - 100,000	13	60,001 - 75,000	7
	18,001 - 22,000	3		100,001 - 115,000	14	75,001 - 80,000	8
	22,001 - 25,000	4		115,001 and over	15	80,001 - 100,000	9
	25,001 - 30,000	5				100,001 and over	10

Table 2: Two-Earner/Two-Job Worksheet

Married Filing Jointly		All Others	
If wages from **HIGHEST** paying job are—	Enter on line 7 above	If wages from **HIGHEST** paying job are—	Enter on line 7 above
$0 - $60,000	$480	$0 - $30,000	$480
60,001 - 110,000	800	30,001 - 70,000	800
110,001 - 160,000	900	70,001 - 140,000	900
160,001 - 280,000	1,060	140,001 - 320,000	1,060
280,001 and over	1,120	320,001 and over	1,120

Privacy Act and Paperwork Reduction Act Notice. We ask for the information on this form to carry out the Internal Revenue laws of the United States. The Internal Revenue Code requires this information under sections 3402(f)(2)(A) and 6109 and their regulations. Failure to provide a properly completed form will result in your being treated as a single person who claims no withholding allowances; providing fraudulent information may also subject you to penalties. Routine uses of this information include giving it to the Department of Justice for civil and criminal litigation, to cities, states, and the District of Columbia for use in administering their tax laws, and using it in the National Directory of New Hires. We may also disclose this information to other countries under a tax treaty, to federal and state agencies to enforce federal nontax criminal laws, or to federal law enforcement and intelligence agencies to combat terrorism.

You are not required to provide the information requested on a form that is subject to the Paperwork Reduction Act unless the form displays a valid OMB control number. Books or records relating to a form or its instructions must be retained as long as their contents may become material in the administration of any Internal Revenue law. Generally, tax returns and return information are confidential, as required by Code section 6103.

The time needed to complete this form will vary depending on individual circumstances. The estimated average time is: Recordkeeping, 45 min.; Learning about the law or the form, 12 min.; Preparing the form, 58 min. If you have comments concerning the accuracy of these time estimates or suggestions for making this form simpler, we would be happy to hear from you. You can write to: Internal Revenue Service, Tax Products Coordinating Committee, SE:W:CAR:MP:T:T:SP, 1111 Constitution Ave. NW, IR-6406, Washington, DC 20224. **Do not** send Form W-4 to this address. Instead, give it to your employer.

Printed on recycled paper

APPENDIX 11
EARNED INCOME CREDIT ADVANCE PAYMENT CERTIFICATE (IRS FORM W-5)

2005 Form W-5

Department of the Treasury
Internal Revenue Service

Instructions

What's New for 2005?

- The definition of a qualifying child has been revised. See *Who Is a Qualifying Child?* below.
- You may elect to include nontaxable combat pay in earned income for purposes of figuring your earned income credit. However, do not include it in earned income for purposes of advance payments of the earned income credit.

Purpose of Form

Use Form W-5 if you are eligible to get part of the earned income credit (EIC) in advance with your pay and choose to do so. See *Who Is Eligible To Get Advance EIC Payments?* below. The amount you can get in advance generally depends on your wages. If you are married, the amount of your advance EIC payments also depends on whether your spouse has filed a Form W-5 with his or her employer. However, your employer cannot give you more than $1,597 throughout 2005 with your pay. You will get the rest of any EIC you are entitled to when you file your tax return and claim the EIC.

If you do not choose to get advance payments, you can still claim the EIC on your 2005 tax return.

What Is the EIC?

The EIC is a credit for certain workers. It reduces the tax you owe. It may give you a refund even if you do not owe any tax.

Who Is Eligible To Get Advance EIC Payments?

You are eligible to get advance EIC payments if **all three** of the following apply.

1. You expect to have at least one qualifying child. If you do not expect to have a qualifying child, you may still be eligible for the EIC, but you cannot receive advance EIC payments. See *Who Is a Qualifying Child?* below.

2. You expect that your 2005 earned income and adjusted gross income (AGI) will each be less than $31,030 ($33,030 if you expect to file a joint return for 2005). Include your spouse's income if you plan to file a joint return. As used on this form, earned income does not include amounts inmates in penal institutions are paid for their work, amounts received as a pension or annuity from a nonqualified deferred compensation plan or a nongovernmental section 457 plan, or nontaxable earned income.

3. You expect to be able to claim the EIC for 2005. To find out if you may be able to claim the EIC, answer the questions on page 2.

How To Get Advance EIC Payments

If you are eligible to get advance EIC payments, fill in the 2005 Form W-5 at the bottom of this page. Then, detach it and give it to your employer. If you get advance payments, you must file a 2005 Form 1040 or 1040A income tax return.

You may have only one Form W-5 in effect at one time. If you and your spouse are both employed, you should file separate Forms W-5.

This Form W-5 expires on December 31, 2005. If you are eligible to get advance EIC payments for 2006, you must file a new Form W-5 next year.

TIP You may be able to get a larger credit when you file your 2005 return. For details, see *Additional Credit* on page 3.

Who Is a Qualifying Child?

A qualifying child is any child who meets all three of the following conditions.

1. The child is:

a. Your son, daughter, adopted child (including a child lawfully placed with you for legal adoption even if the adoption is not final), stepchild, or a descendant of any of them (for example, your grandchild), or

b. Your brother, sister, stepbrother, stepsister, or a descendant of any of them (for example, your niece or nephew), or

c. A foster child (any child placed with you by an authorized placement agency or by judgment, decree, or other order of any court of competent jurisdiction).

(continued on page 3)

Give the bottom part to your employer; keep the top part for your records.

Detach here

Form **W-5** — **Earned Income Credit Advance Payment Certificate** — OMB No. 1545-1342 — **2005**

Department of the Treasury
Internal Revenue Service

► **Use the current year's certificate only.**
► **Give this certificate to your employer.**
► **This certificate expires on December 31, 2005.**

Print or type your full name | **Your social security number**

Note: *If you get advance payments of the earned income credit for 2005, you* **must** *file a 2005 federal income tax return. To get advance payments, you* **must** *have a qualifying child and your filing status must be any status* **except** *married filing a separate return.*

1 I expect to have a qualifying child and be able to claim the earned income credit for 2005, I do not have another Form W-5 in effect with any other current employer, and I choose to get advance EIC payments ☐ **Yes** ☐ **No**

2 Check the box that shows your expected filing status for 2005:
☐ Single, head of household, or qualifying widow(er) ☐ Married filing jointly

3 If you are married, does your spouse have a Form W-5 in effect for 2005 with any employer? ☐ **Yes** ☐ **No**

Under penalties of perjury, I declare that the information I have furnished above is, to the best of my knowledge, true, correct, and complete.

Signature ► **Date ►**

Cat. No. 10227P

Form W-5 (2005) Page **2**

Questions To See if You May Be Able To Claim the EIC for 2005

CAUTION: You **cannot** claim the EIC if you plan to file either Form 2555 or Form 2555-EZ (relating to foreign earned income) for 2005. You also **cannot** claim the EIC if you are a nonresident alien for any part of 2005 unless you are married to a U.S. citizen or resident, file a joint return, and elect to be taxed as a resident alien for all of 2005.

1 Do you expect to have a qualifying child? Read *Who Is a Qualifying Child?* that starts on page 1 before you answer this question. If the child is married, be sure you also read *Married child* on page 3.

☐ **No.** STOP You may be able to claim the EIC but you **cannot** get advance EIC payments.
☐ **Yes.** *Continue.*

CAUTION: If the child meets the conditions to be a qualifying child for both you and another person, see *Qualifying child of more than one person* on page 3.

2 Do you expect your 2005 filing status to be married filing a separate return?

☐ **Yes.** STOP You **cannot** claim the EIC.
☐ **No.** *Continue.*

TIP: If you expect to file a joint return for 2005, include your spouse's income when answering questions 3 and 4.

3 Do you expect that your 2005 earned income and AGI will each be less than: $31,030 ($33,030 if married filing jointly) if you expect to have 1 qualifying child; $35,263 ($37,263 if married filing jointly) if you expect to have 2 or more qualifying children?

☐ **No.** STOP You **cannot** claim the EIC.
☐ **Yes.** *Continue.* But remember, you **cannot** get advance EIC payments if you expect your 2005 earned income or AGI will be $31,030 ($33,030 or more if married filing jointly) or more.

4 Do you expect that your 2005 investment income will be more than $2,700? For most people, investment income is the total of their taxable interest, ordinary dividends, capital gain distributions, and tax-exempt interest. However, if you plan to file a 2005 Form 1040, see the 2004 Form 1040 instructions to figure your investment income.

☐ **Yes.** STOP You **cannot** claim the EIC.
☐ **No.** *Continue.*

5 Do you expect that you, or your spouse if filing a joint return, will be a qualifying child of another person for 2005?
☐ **No.** You may be able to claim the EIC.
☐ **Yes.** You **cannot** claim the EIC.

Form W-5 (2005) Page **3**

2. At the end of 2005, the child is under age 19, or under age 24 and a student, or any age and permanently and totally disabled. A student is a child who during any 5 months of 2004 (a) was enrolled as a full-time student at a school or (b) took a full-time, on-farm training course given by a school or a state, county, or local government agency. A school includes a technical, trade, or mechanical school. It does not include an on-the-job training course, correspondence school, or night school.

3. The child lives with you in the United States for over half of 2005. But you do not have to meet this condition if (a) the child was born or died during the year and your home was this child's home for the entire time he or she was alive in 2005, or (b) the child is presumed by law enforcement authorities to have been kidnapped by someone who is not a family member and the child lived with you for over half of the part of the year before he or she was kidnapped.

Note. Temporary absences, such as for school, vacation, medical care, or detention in a juvenile facility, count as time lived at home. Members of the military on extended active duty outside the United States are considered to be living in the United States.

Married child. A child who is married at the end of 2005 is a qualifying child only if:

1. You may claim him or her as your dependent, or

2. You are the custodial parent and would be able to claim the child as your dependent, but the noncustodial parent claims the child as a dependent because:

a. You signed Form 8332, Release of Claim to Exemption for Child of Divorced or Separated Parents, or a similar statement, agreeing not to claim the child for 2005, or

b. You have a pre-1985 divorce decree or separation agreement that allows the noncustodial parent to claim the child and he or she gives at least $600 for the child's support in 2005.

Other rules may apply. See Pub. 501, Exemptions, Standard Deduction, and Filing Information, for more information on children of divorced or separated parents.

Qualifying child of more than one person. If the child meets the conditions to be a qualifying child of more than one person, only one person may treat that child as a qualifying child for 2005. If more than one person actually claims the credit based on the same qualifying child, the IRS will apply the "tie-breaker" rules to determine which person can treat the child as a qualifying child for 2005. The "tie-breaker" rules are in Table 2, When More Than One Person Claims EIC Using Same Child, in Pub. 596, Earned Income Credit (EIC).

Caution. A qualifying child whom you use to claim the EIC must have a valid social security number unless he or she was born and died in 2005.

What if My Situation Changes?

If your situation changes after you give Form W-5 to your employer, you will probably need to file a new Form W-5. For example, you must file a new Form W-5 if any of the following applies for 2005.

- You no longer expect to have a qualifying child. Check "No" on line 1 of your new Form W-5.
- You no longer expect to be able to claim the EIC for 2005. Check "No" on line 1 of your new Form W-5.
- You no longer want advance payments. Check "No" on line 1 of your new Form W-5.
- Your spouse files Form W-5 with his or her employer. Check "Yes" on line 3 of your new Form W-5.

Note. If you get advance EIC payments and find you are not eligible for the EIC, you must pay back these payments when you file your 2005 federal income tax return.

Additional Information

How To Claim the EIC

If you are eligible, claim the EIC on your 2005 tax return. See your 2005 tax return instruction booklet.

Additional Credit

You may be able to claim a larger credit when you file your 2005 Form 1040 or Form 1040A because your employer cannot give you more than $1,597 throughout the year with your pay. You may also be able to claim a larger credit if you have more than one qualifying child. But you must file your 2005 tax return to claim any additional credit.

Privacy Act and Paperwork Reduction Act Notice. We ask for the information on this form to carry out the Internal Revenue laws of the United States. Internal Revenue Code sections 3507 and 6109 and their regulations require you to provide the information requested on Form W-5 and to give it to your employer if you want advance payment of the EIC. As provided by law, we may give the information to the Department of Justice and other federal agencies. In addition, we may give it to cities, states, and the District of Columbia so they may carry out their tax laws. We may also disclose this information to other countries under a tax treaty, to federal and state agencies to enforce federal nontax criminal laws, or to federal law enforcement and intelligence agencies to combat terrorism. Failure to provide the requested information may prevent your employer from processing this form; providing false information may subject you to penalties.

You are not required to provide the information requested on a form that is subject to the Paperwork Reduction Act unless the form displays a valid OMB control number. Books or records relating to a form or its instructions must be retained as long as their contents may become material in the administration of any Internal Revenue law. Generally, tax returns and return information are confidential, as required by Code section 6103.

The time needed to complete this form will vary depending on individual circumstances. The estimated average time is: Recordkeeping, 6 min.; Learning about the law or the form, 12 min.; and Preparing the form, 25 min.

We welcome comments on forms. If you have comments concerning the accuracy of these time estimates or suggestions for making this form simpler, we would be happy to hear from you. You can write to Internal Revenue Service, Tax Products Coordinating Committee, SE:W:CAR:MP:T:T:SP, 1111 Constitution Ave. NW, IR-6406, Washington, DC 20224. Do not send Form W-5 to this address. Instead, give it to your employer.

APPENDIX 12
IRS SCHEDULE H

SCHEDULE H (Form 1040)

Department of the Treasury
Internal Revenue Service (99)

Household Employment Taxes

(For Social Security, Medicare, Withheld Income, and Federal Unemployment (FUTA) Taxes)

► **Attach to Form 1040, 1040NR, 1040-SS, or 1041.**

► **See separate instructions.**

OMB No. 1545-0074

2004

Attachment Sequence No. **44**

Name of employer

Social security number

Employer identification number

A Did you pay **any one** household employee cash wages of $1,400 or more in 2004? (If any household employee was your spouse, your child under age 21, your parent, or anyone under age 18, see the line A instructions on page H-3 before you answer this question.)

☐ **Yes.** Skip lines B and C and go to line 1.
☐ **No.** Go to line B.

B Did you withhold Federal income tax during 2004 for any household employee?

☐ **Yes.** Skip line C and go to line 5.
☐ **No.** Go to line C.

C Did you pay **total** cash wages of $1,000 or more in **any** calendar **quarter** of 2003 or 2004 to **all** household employees? (**Do not** count cash wages paid in 2003 or 2004 to your spouse, your child under age 21, or your parent.)

☐ **No. Stop.** Do not file this schedule.
☐ **Yes.** Skip lines 1-9 and go to line 10 on the back. (Calendar year taxpayers having no household employees in 2004 **do not** have to complete this form for 2004.)

Part I **Social Security, Medicare, and Income Taxes**

1	Total cash wages subject to social security taxes (see page H-3)	1		
2	Social security taxes. Multiply line 1 by 12.4% (.124)		2	
3	Total cash wages subject to Medicare taxes (see page H-3)	3		
4	Medicare taxes. Multiply line 3 by 2.9% (.029)		4	
5	Federal income tax withheld, if any		5	
6	**Total social security, Medicare, and income taxes** (add lines 2, 4, and 5)		6	
7	Advance earned income credit (EIC) payments, if any		7	
8	**Net taxes** (subtract line 7 from line 6)		8	

9 Did you pay **total** cash wages of $1,000 or more in **any** calendar **quarter** of 2003 or 2004 to household employees? (**Do not** count cash wages paid in 2003 or 2004 to your spouse, your child under age 21, or your parent.)

☐ **No. Stop.** Enter the amount from line 8 above on Form 1040, line 61. If you are not required to file Form 1040, see the line 9 instructions on page H-4.

☐ **Yes.** Go to line 10 on the back.

For Paperwork Reduction Act Notice, see Form 1040 instructions. Cat. No. 12187K **Schedule H (Form 1040) 2004**

Schedule H (Form 1040) 2004 Page 2

Part II Federal Unemployment (FUTA) Tax

			Yes	No
10	Did you pay unemployment contributions to only one state? (If you paid contributions to New York State, check "No.")	10		
11	Did you pay all state unemployment contributions for 2004 by April 15, 2005? Fiscal year filers, see page H-4	11		
12	Were all wages that are taxable for FUTA tax also taxable for your state's unemployment tax?	12		

Next: If you checked the **"Yes"** box on **all** the lines above, complete Section A.
If you checked the **"No"** box on **any** of the lines above, skip Section A and complete Section B.

Section A

13 Name of the state where you paid unemployment contributions ▶ ______

14 State reporting number as shown on state unemployment tax return ▶ ______

15 Contributions paid to your state unemployment fund (see page H-4) | 15 | ______

16 Total cash wages subject to FUTA tax (see page H-4) | 16 | ______

17 **FUTA tax.** Multiply line 16 by .008. Enter the result here, skip Section B, and go to line 26 | 17 | ______

Section B

18 Complete all columns below that apply (if you need more space, see page H-4):

(a) Name of state	(b) State reporting number as shown on state unemployment tax return	(c) Taxable wages (as defined in state act)	(d) State experience rate period From	(d) State experience rate period To	(e) State experience rate	(f) Multiply col. (c) by .054	(g) Multiply col. (c) by col. (e)	(h) Subtract col. (g) from col. (f). If zero or less, enter -0-.	(i) Contributions paid to state unemployment fund

19 Totals | 19 | ______

20 Add columns (h) and (i) of line 19 | 20 | ______

21 Total cash wages subject to FUTA tax (see the line 16 instructions on page H-4) | 21 | ______

22 Multiply line 21 by 6.2% (.062) | 22 | ______

23 Multiply line 21 by 5.4% (.054) | 23 | ______

24 Enter the **smaller** of line 20 or line 23 | 24 | ______
(New York State employers must use the worksheet in the separate instructions and check here) ☐

25 **FUTA tax.** Subtract line 24 from line 22. Enter the result here and go to line 26 | 25 | ______

Part III Total Household Employment Taxes

26 Enter the amount from line 8 | 26 | ______

27 Add line 17 (or line 25) and line 26 | 27 | ______

28 Are you required to file Form 1040?

☐ **Yes. Stop.** Enter the amount from line 27 above on Form 1040, line 61. **Do not** complete Part IV below.

☐ **No.** You may have to complete Part IV. See page H-4 for details.

Part IV Address and Signature—Complete this part **only** if required. See the line 28 instructions on page H-4.

Address (number and street) or P.O. box if mail is not delivered to street address | Apt., room, or suite no.

City, town or post office, state, and ZIP code

Under penalties of perjury, I declare that I have examined this schedule, including accompanying statements, and to the best of my knowledge and belief, it is true, correct, and complete. No part of any payment made to a state unemployment fund claimed as a credit was, or is to be, deducted from the payments to employees.

▶ ______ Employer's signature ▶ ______ Date

Schedule H (Form 1040) 2004

APPENDIX 13
APPLICATION FOR EMPLOYER IDENTIFICATION NUMBER (IRS FORM SS-4)

Form **SS-4**
(Rev. December 2001)
Department of the Treasury
Internal Revenue Service

Application for Employer Identification Number
(For use by employers, corporations, partnerships, trusts, estates, churches, government agencies, Indian tribal entities, certain individuals, and others.)
▶ **See separate instructions for each line.** ▶ **Keep a copy for your records.**

EIN

OMB No. 1545-0003

Type or print clearly.

1 Legal name of entity (or individual) for whom the EIN is being requested

2 Trade name of business (if different from name on line 1) | 3 Executor, trustee, "care of" name

4a Mailing address (room, apt., suite no. and street, or P.O. box) | 5a Street address (if different) (Do not enter a P.O. box.)

4b City, state, and ZIP code | 5b City, state, and ZIP code

6 County and state where principal business is located

7a Name of principal officer, general partner, grantor, owner, or trustor | 7b SSN, ITIN, or EIN

8a Type of entity (check only one box)
- Sole proprietor (SSN)
- Partnership
- Corporation (enter form number to be filed) ▶
- Personal service corp.
- Church or church-controlled organization
- Other nonprofit organization (specify) ▶
- Other (specify) ▶
- Estate (SSN of decedent)
- Plan administrator (SSN)
- Trust (SSN of grantor)
- National Guard
- State/local government
- Farmers' cooperative
- Federal government/military
- REMIC
- Indian tribal governments/enterprises

Group Exemption Number (GEN) ▶

8b If a corporation, name the state or foreign country (if applicable) where incorporated | State | Foreign country

9 Reason for applying (check only one box)
- Started new business (specify type) ▶
- Hired employees (Check the box and see line 12.)
- Compliance with IRS withholding regulations
- Other (specify) ▶
- Banking purpose (specify purpose) ▶
- Changed type of organization (specify new type) ▶
- Purchased going business
- Created a trust (specify type) ▶
- Created a pension plan (specify type) ▶

10 Date business started or acquired (month, day, year) | **11** Closing month of accounting year

12 First date wages or annuities were paid or will be paid (month, day, year). **Note:** If applicant is a withholding agent, enter date income will first be paid to nonresident alien. (month, day, year) ▶

13 Highest number of employees expected in the next 12 months. **Note:** If the applicant does not expect to have any employees during the period, enter "-0-." ▶

Agricultural	Household	Other

14 Check **one** box that best describes the principal activity of your business.
- Health care & social assistance
- Wholesale–agent/broker
- Construction
- Rental & leasing
- Transportation & warehousing
- Accommodation & food service
- Wholesale–other
- Retail
- Real estate
- Manufacturing
- Finance & insurance
- Other (specify)

15 Indicate principal line of merchandise sold; specific construction work done; products produced; or services provided.

16a Has the applicant ever applied for an employer identification number for this or any other business? — **Yes** — **No**
Note: If "Yes," please complete lines 16b and 16c.

16b If you checked "Yes" on line 16a, give applicant's legal name and trade name shown on prior application if different from line 1 or 2 above.
Legal name ▶ Trade name ▶

16c Approximate date when, and city and state where, the application was filed. Enter previous employer identification number if known.
Approximate date when filed (mo., day, year) | City and state where filed | Previous EIN

Third Party Designee
Complete this section **only** if you want to authorize the named individual to receive the entity's EIN and answer questions about the completion of this form.
Designee's name | Designee's telephone number (include area code) ( )
Address and ZIP code | Designee's fax number (include area code) ( )

Under penalties of perjury, I declare that I have examined this application, and to the best of my knowledge and belief, it is true, correct, and complete.
Applicant's telephone number (include area code) ( )
Name and title (type or print clearly) ▶
Applicant's fax number (include area code) ( )
Signature ▶ Date ▶

For Privacy Act and Paperwork Reduction Act Notice, see separate instructions. Cat. No. 16055N Form **SS-4** (Rev. 12-2001)

Form SS-4 (Rev. 12-2001) Page **2**

Do I Need an EIN?

File Form SS-4 if the applicant entity does not already have an EIN but is required to show an EIN on any return, statement, or other document.[1] **See also the separate instructions for each line on Form SS-4.**

IF the applicant...	**AND...**	**THEN...**
Started a new business	Does not currently have (nor expect to have) employees	Complete lines 1, 2, 4a–6, 8a, and 9–16c.
Hired (or will hire) employees, including household employees	Does not already have an EIN	Complete lines 1, 2, 4a–6, 7a–b (if applicable), 8a, 8b (if applicable), and 9–16c.
Opened a bank account	Needs an EIN for banking purposes only	Complete lines 1–5b, 7a–b (if applicable), 8a, 9, and 16a–c.
Changed type of organization	Either the legal character of the organization or its ownership changed (e.g., you incorporate a sole proprietorship or form a partnership)[2]	Complete lines 1–16c (as applicable).
Purchased a going business[3]	Does not already have an EIN	Complete lines 1–16c (as applicable).
Created a trust	The trust is other than a grantor trust or an IRA trust[4]	Complete lines 1–16c (as applicable).
Created a pension plan as a plan administrator[5]	Needs an EIN for reporting purposes	Complete lines 1, 2, 4a–6, 8a, 9, and 16a–c.
Is a foreign person needing an EIN to comply with IRS withholding regulations	Needs an EIN to complete a Form W-8 (other than Form W-8ECI), avoid withholding on portfolio assets, or claim tax treaty benefits[6]	Complete lines 1–5b, 7a–b (SSN or ITIN optional), 8a–9, and 16a–c.
Is administering an estate	Needs an EIN to report estate income on Form 1041	Complete lines 1, 3, 4a–b, 8a, 9, and 16a–c.
Is a withholding agent for taxes on non-wage income paid to an alien (i.e., individual, corporation, or partnership, etc.)	Is an agent, broker, fiduciary, manager, tenant, or spouse who is required to file **Form 1042,** Annual Withholding Tax Return for U.S. Source Income of Foreign Persons	Complete lines 1, 2, 3 (if applicable), 4a–5b, 7a–b (if applicable), 8a, 9, and 16a–c.
Is a state or local agency	Serves as a tax reporting agent for public assistance recipients under Rev. Proc. 80-4, 1980-1 C.B. 581[7]	Complete lines 1, 2, 4a–5b, 8a, 9, and 16a–c.
Is a single-member LLC	Needs an EIN to file **Form 8832,** Classification Election, for filing employment tax returns, **or** for state reporting purposes[8]	Complete lines 1–16c (as applicable).
Is an S corporation	Needs an EIN to file **Form 2553,** Election by a Small Business Corporation[9]	Complete lines 1–16c (as applicable).

[1] For example, a sole proprietorship or self-employed farmer who establishes a qualified retirement plan, or is required to file excise, employment, alcohol, tobacco, or firearms returns, must have an EIN. **A partnership, corporation, REMIC (real estate mortgage investment conduit), nonprofit organization (church, club, etc.), or farmers' cooperative must use an EIN for any tax-related purpose even if the entity does not have employees.**

[2] However, **do not** apply for a new EIN if the existing entity only **(a)** changed its business name, **(b)** elected on Form 8832 to change the way it is taxed (or is covered by the default rules), or **(c)** terminated its partnership status because at least 50% of the total interests in partnership capital and profits were sold or exchanged within a 12-month period. (The EIN of the terminated partnership should continue to be used. See Regulations section 301.6109-1(d)(2)(iii).)

[3] Do not use the EIN of the prior business unless you became the "owner" of a corporation by acquiring its stock.

[4] However, IRA trusts that are required to file **Form 990-T,** Exempt Organization Business Income Tax Return, must have an EIN.

[5] A plan administrator is the person or group of persons specified as the administrator by the instrument under which the plan is operated.

[6] Entities applying to be a Qualified Intermediary (QI) need a QI-EIN even if they already have an EIN. **See Rev. Proc. 2000-12.**

[7] See also Household employer on page 4. (**Note:** State or local agencies may need an EIN for other reasons, e.g., hired employees.)

[8] Most LLCs **do not** need to file Form 8832. See **Limited liability company (LLC)** on page 4 for details on completing Form SS-4 for an LLC.

[9] An existing corporation that is electing or revoking S corporation status should use its previously-assigned EIN.

APPENDIX 14: DIRECTORY OF IRS MAILING ADDRESSES AND FAX-TIN NUMBERS FOR FILING IRS FORM SS-4

STATES COVERED	MAILING ADDRESS	FAX-TIN NUMBER
Connecticut, Delaware, District of Columbia, Florida, Georgia, Maine, Maryland, Massachusetts, New Hampshire, New Jersey, New York, North Carolina, Ohio, Pennsylvania, Rhode Island, South Carolina, Vermont, Virginia, West Virginia	Internal Revenue Service Center EIN Operation P.O. Box 9003 Holtsville, NY, 11742-9003	631-447-8960
Illinois, Indiana, Kentucky, Michigan	Internal Revenue Service Center EIN Operation Cincinnati, OH 45999	859-669-5760
Alabama, Alaska, Arizona, Arkansas, California, Colorado, Hawaii, Idaho, Iowa, Kansas, Louisiana, Minnesota, Mississippi, Missouri, Montana, Nebraska, Nevada, New Mexico, North Dakota, Oklahoma, Oregon, Puerto Rico, South Dakota, Tennessee, Texas, Utah, Washington, Wisconsin, Wyoming	Internal Revenue Service Center EIN Operation Philadelphia, PA 19255	215-516-3990
No legal residence, place of business, or principal office or agency in any state	Internal Revenue Service Center EIN Operation Philadelphia, PA 19255	215-516-3990

APPENDIX 15
EMPLOYEE WAGE AND TAX STATEMENT (IRS FORM W-2)

Attention:

This form is provided for informational purposes only. Copy A appears in red, similar to the official printed IRS form. But do not file Copy A downloaded from this website with the SSA. A penalty of $50 per information return may be imposed for filing such forms that cannot be scanned.

To order official IRS forms, call 1-800-TAX-FORMS (1-800-829-3676) or order online at Forms and Publications By U.S. Mail.

You may file Forms W-2 and W-3 electronically on the SSA's website at Employer Reporting Instructions & Information. You can create fill-in versions of Forms W-2 and W-3 for filing with the SSA. You may also print out copies for filing with state or local governments, distribution to your employees, and for your records.

W-2 / W-3 Cover page

a Control number — 22222 — Void ☐ — **For Official Use Only ▶** **OMB No. 1545-0008**

b Employer identification number (EIN)	1 Wages, tips, other compensation	2 Federal income tax withheld
c Employer's name, address, and ZIP code	3 Social security wages	4 Social security tax withheld
	5 Medicare wages and tips	6 Medicare tax withheld
	7 Social security tips	8 Allocated tips
d Employee's social security number	9 Advance EIC payment	10 Dependent care benefits
e Employee's first name and initial — Last name	11 Nonqualified plans	12a See instructions for box 12 (Code)
	13 Statutory employee ☐ Retirement plan ☐ Third-party sick pay ☐	12b (Code)
	14 Other	12c (Code)
		12d (Code)
f Employee's address and ZIP code		

15 State	Employer's state ID number	16 State wages, tips, etc.	17 State income tax	18 Local wages, tips, etc.	19 Local income tax	20 Locality name

Form **W-2** **Wage and Tax Statement** **2005**

Department of the Treasury—Internal Revenue Service

For Privacy Act and Paperwork Reduction Act Notice, see back of Copy D.

Copy A For Social Security Administration — Send this entire page with Form W-3 to the Social Security Administration; photocopies are **not** acceptable.

Cat. No. 10134D

Do Not Cut, Fold, or Staple Forms on This Page — Do Not Cut, Fold, or Staple Forms on This Page

a Control number	22222	OMB No. 1545-0008	
b Employer identification number (EIN)		1 Wages, tips, other compensation	2 Federal income tax withheld
c Employer's name, address, and ZIP code		3 Social security wages	4 Social security tax withheld
		5 Medicare wages and tips	6 Medicare tax withheld
		7 Social security tips	8 Allocated tips
d Employee's social security number		9 Advance EIC payment	10 Dependent care benefits
e Employee's first name and initial	Last name	11 Nonqualified plans	12a Code
		13 Statutory employee ☐ Retirement plan ☐ Third-party sick pay ☐	12b Code
		14 Other	12c Code
			12d Code
f Employee's address and ZIP code			

15 State	Employer's state ID number	16 State wages, tips, etc.	17 State income tax	18 Local wages, tips, etc.	19 Local income tax	20 Locality name

Form **W-2** **Wage and Tax Statement** 2005 Department of the Treasury—Internal Revenue Service

Copy 1—For State, City, or Local Tax Department

a Control number	OMB No. 1545-0008	Safe, accurate, FAST! Use IRS e-file	Visit the IRS website at *www.irs.gov/efile*.
b Employer identification number (EIN)		1 Wages, tips, other compensation	2 Federal income tax withheld
c Employer's name, address, and ZIP code		3 Social security wages	4 Social security tax withheld
		5 Medicare wages and tips	6 Medicare tax withheld
		7 Social security tips	8 Allocated tips
d Employee's social security number		9 Advance EIC payment	10 Dependent care benefits
e Employee's first name and initial	Last name	11 Nonqualified plans	12a See instructions for box 12 Code
		13 Statutory employee / Retirement plan / Third-party sick pay	12b Code
		14 Other	12c Code
			12d Code
f Employee's address and ZIP code			

15 State	Employer's state ID number	16 State wages, tips, etc.	17 State income tax	18 Local wages, tips, etc.	19 Local income tax	20 Locality name

Form **W-2** **Wage and Tax Statement** **2005** Department of the Treasury—Internal Revenue Service

Copy B—To Be Filed With Employee's FEDERAL Tax Return.
This information is being furnished to the Internal Revenue Service.

Notice to Employee

Refund. Even if you do not have to file a tax return, you should file to get a refund if box 2 shows federal income tax withheld or if you can take the earned income credit.

Earned income credit (EIC). You must file a tax return if any amount is shown in box 9.

You may be able to take the EIC for 2005 if: **(a)** you do not have a qualifying child and you earned less than $11,750 ($13,750 if married filing jointly), **(b)** you have one qualifying child and you earned less than $31,030 ($33,030 if married filing jointly), or **(c)** you have more than one qualifying child and you earned less than $35,263 ($37,263 if married filing jointly). You and any qualifying children must have valid social security numbers (SSNs). You cannot take the EIC if your investment income is more than $2,700. **Any EIC that is more than your tax liability is refunded to you, but only if you file a tax return.** If you have at least one qualifying child, you may get as much as $1,597 of the EIC in advance by completing Form W-5, Earned Income Credit Advance Payment Certificate, and giving it to your employer.

Clergy and religious workers. If you are not subject to social security and Medicare taxes, see Publication 517, Social Security and Other Information for Members of the Clergy and Religious Workers.

Corrections. If your name, SSN, or address is incorrect, correct Copies B, C, and 2 and ask your employer to correct your employment record. Be sure to ask the employer to file Form W-2c, Corrected Wage and Tax Statement, with the Social Security Administration (SSA) to correct any name, SSN, or money amount error reported to the SSA on Form W-2. If your name and SSN are correct but are not the same as shown on your social security card, you should ask for a new card at any SSA office or call 1-800-772-1213.

Credit for excess taxes. If you had more than one employer in 2005 and more than $5,580.00 in social security and/or Tier I railroad retirement (RRTA) taxes were withheld, you may be able to claim a credit for the excess against your federal income tax. If you had more than one railroad employer and more than $2,943.60 in Tier II RRTA tax was withheld, you also may be able to claim a credit. See your Form 1040 or Form 1040A instructions and Publication 505, Tax Withholding and Estimated Tax.

(Also see *Instructions* on back of Copy C.)

a Control number

OMB No. 1545-0008

This information is being furnished to the Internal Revenue Service. If you are required to file a tax return, a negligence penalty or other sanction may be imposed on you if this income is taxable and you fail to report it.

b Employer identification number (EIN)

1 Wages, tips, other compensation

2 Federal income tax withheld

c Employer's name, address, and ZIP code

3 Social security wages

4 Social security tax withheld

5 Medicare wages and tips

6 Medicare tax withheld

7 Social security tips

8 Allocated tips

d Employee's social security number

9 Advance EIC payment

10 Dependent care benefits

e Employee's first name and initial Last name

11 Nonqualified plans

12a See instructions for box 12

13 Statutory employee ☐ Retirement plan ☐ Third-party sick pay ☐

12b

14 Other

12c

12d

f Employee's address and ZIP code

15 State	Employer's state ID number	16 State wages, tips, etc.	17 State income tax	18 Local wages, tips, etc.	19 Local income tax	20 Locality name

Form **W-2** **Wage and Tax Statement**

2005

Department of the Treasury—Internal Revenue Service

Copy C—For EMPLOYEE'S RECORDS. (See Notice to Employee on back of Copy B.)

Safe, accurate, FAST! Use IRS e-file

Instructions *(Also see Notice to Employee, on back of Copy B.)*

Box 1. Enter this amount on the wages line of your tax return.

Box 2. Enter this amount on the federal income tax withheld line of your tax return.

Box 8. This amount is **not** included in boxes 1, 3, 5, or 7. For information on how to report tips on your tax return, see your Form 1040 instructions.

Box 9. Enter this amount on the advance earned income credit payments line of your Form 1040 or Form 1040A.

Box 10. This amount is the total dependent care benefits that your employer paid to you or incurred on your behalf (including amounts from a section 125 (cafeteria) plan). Any amount over $5,000 also is included in box 1. You **must** complete Schedule 2 (Form 1040A) or Form 2441, Child and Dependent Care Expenses, to compute any taxable and nontaxable amounts.

Box 11. This amount is: **(a)** reported in box 1 if it is a distribution made to you from a nonqualified deferred compensation or nongovernmental section 457(b) plan or **(b)** included in box 3 and/or 5 if it is a prior year deferral under a nonqualified or section 457(b) plan that became taxable for social security and Medicare taxes this year because there is no longer a substantial risk of forfeiture of your right to the deferred amount.

Box 12. The following list explains the codes shown in box 12. You may need this information to complete your tax return. Elective deferrals (codes D, E, F, and S) under all plans are generally limited to a total of $14,000 ($17,000 for section 403(b) plans if you qualify for the 15-year rule explained in Pub. 571). Deferrals under code G are limited to $14,000. Deferrals under code H are limited to $7,000. However, if you were at least age 50 in 2005, your employer may have allowed an additional deferral of up to $4,000 ($2,000 for section 401(k)(11) and 408(p) SIMPLE plans). This additional deferral amount is not subject to the overall limit on elective deferrals. For code G, the limit on elective deferrals may be higher for the last three years before you reach retirement age. Contact your plan administrator for more information. Amounts in excess of the overall elective deferral limit must be included in income. See the "Wages, Salaries, Tips, etc." line instructions for Form 1040.

Note. *If a year follows code D, E, F, G, H, or S, you made a make-up pension contribution for a prior year(s) when you were in military service. To figure whether you made excess deferrals, consider these amounts for the year shown, not the current year. If no year is shown, the contributions are for the current year.*

A—Uncollected social security or RRTA tax on tips. Include this tax on Form 1040. See "Total Tax" in the Form 1040 instructions.

B—Uncollected Medicare tax on tips. Include this tax on Form 1040. See "Total Tax" in the Form 1040 instructions.

C—Taxable cost of group-term life insurance over $50,000 (included in boxes 1, 3 (up to social security wage base), and 5)

D—Elective deferrals to a section 401(k) cash or deferred arrangement. Also includes deferrals under a SIMPLE retirement account that is part of a section 401(k) arrangement.

E—Elective deferrals under a section 403(b) salary reduction agreement

F—Elective deferrals under a section 408(k)(6) salary reduction SEP

G—Elective deferrals and employer contributions (including nonelective deferrals) to a section 457(b) deferred compensation plan

H—Elective deferrals to a section 501(c)(18)(D) tax-exempt organization plan. See "Adjusted Gross Income" in the Form 1040 instructions for how to deduct.

J—Nontaxable sick pay (information only, not included in boxes 1, 3, or 5)

K—20% excise tax on excess golden parachute payments. See "Total Tax" in the Form 1040 instructions.

L—Substantiated employee business expense reimbursements (nontaxable)

M—Uncollected social security or RRTA tax on taxable cost of group-term life insurance over $50,000 (former employees only). See "Total Tax" in the Form 1040 instructions.

N—Uncollected Medicare tax on taxable cost of group-term life insurance over $50,000 (former employees only). See "Total Tax" in the Form 1040 instructions.

P—Excludable moving expense reimbursements paid directly to employee (not included in boxes 1, 3, or 5)

Q—Nontaxable combat pay. See the instructions for Form 1040 or Form 1040A for details on reporting this amount.

R—Employer contributions to your Archer MSA. Report on Form 8853, Archer MSAs and Long-Term Care Insurance Contracts.

S—Employee salary reduction contributions under a section 408(p) SIMPLE (not included in box 1)

T—Adoption benefits (not included in box 1). You **must** complete Form 8839, Qualified Adoption Expenses, to compute any taxable and nontaxable amounts.

V—Income from exercise of nonstatutory stock option(s) (included in boxes 1, 3 (up to social security wage base), and 5)

W—Employer contributions to your Health Savings Account. Report on Form 8889, Health Savings Accounts (HSAs).

Y—Deferrals under a section 409A nonqualified deferred compensation plan.

Z—Income under section 409A on a nonqualified deferred compensation plan. This amount is also included in box 1. It is subject to an additional 20% tax plus interest. See "Total Tax" in the Form 1040 instructions.

Box 13. If the "Retirement plan" box is checked, special limits may apply to the amount of traditional IRA contributions that you may deduct.

Note: *Keep **Copy C** of Form W-2 for at least 3 years after the due date for filing your income tax return. However, to help **protect your social security benefits,** keep Copy C until you begin receiving social security benefits, just in case there is a question about your work record and/or earnings in a particular year. Review the information shown on your annual (for workers over 25) Social Security Statement.*

a Control number

OMB No. 1545-0008

b Employer identification number (EIN)

1 Wages, tips, other compensation

2 Federal income tax withheld

c Employer's name, address, and ZIP code

3 Social security wages

4 Social security tax withheld

5 Medicare wages and tips

6 Medicare tax withheld

7 Social security tips

8 Allocated tips

d Employee's social security number

9 Advance EIC payment

10 Dependent care benefits

e Employee's first name and initial Last name

11 Nonqualified plans

12a Code

13 Statutory employee Retirement plan Third-party sick pay

12b Code

14 Other

12c Code

12d Code

f Employee's address and ZIP code

15 State Employer's state ID number	16 State wages, tips, etc.	17 State income tax	18 Local wages, tips, etc.	19 Local income tax	20 Locality name

Form **W-2** **Wage and Tax Statement** 2005

Department of the Treasury—Internal Revenue Service

Copy 2—To Be Filed With Employee's State, City, or Local Income Tax Return.

a Control number

Void ☐

OMB No. 1545-0008

b Employer identification number (EIN)

1 Wages, tips, other compensation

2 Federal income tax withheld

c Employer's name, address, and ZIP code

3 Social security wages

4 Social security tax withheld

5 Medicare wages and tips

6 Medicare tax withheld

7 Social security tips

8 Allocated tips

d Employee's social security number

9 Advance EIC payment

10 Dependent care benefits

e Employee's first name and initial   Last name

11 Nonqualified plans

12a See instructions for box 12 (Code)

13 Statutory employee ☐ Retirement plan ☐ Third-party sick pay ☐

12b Code

14 Other

12c Code

12d Code

f Employee's address and ZIP code

15 State	Employer's state ID number	16 State wages, tips, etc.	17 State income tax	18 Local wages, tips, etc.	19 Local income tax	20 Locality name

Form **W-2** **Wage and Tax Statement**

2005

Department of the Treasury—Internal Revenue Service

For Privacy Act and Paperwork Reduction Act Notice, see back of Copy D.

Copy D—For Employer.

Employers, Please Note—

Specific information needed to complete Form W-2 is given in a separate booklet titled 2005 Instructions for Forms W-2 and W-3. You can order those instructions and additional forms by calling 1-800-TAX-FORM (1-800-829-3676). You can also get forms and instructions from the IRS website at *www.irs.gov.*

Caution. *Because the SSA processes paper forms by machine, you cannot file with the SSA Forms W-2 and W-3 that you print from the IRS website. Instead, you can use the SSA website at www.socialsecurity.gov/employer to create and file electronically "fill-in" versions of Forms W-2 and W-3.*

Due dates. Furnish Copies B, C, and 2 to the employee generally by January 31, 2006.

File Copy A with the SSA generally by February 28, 2006. Send all Copies A with Form W-3, Transmittal of Wage and Tax Statements. However, if you file electronically (not by magnetic media), the due date is March 31, 2006.

Privacy Act and Paperwork Reduction Act Notice. We ask for the information on Forms W-2 and W-3 to carry out the Internal Revenue laws of the United States. We need it to figure and collect the right amount of tax. Section 6051 and its regulations require you to furnish wage and tax statements to employees and to the Social Security Administration. Section 6109 requires you to provide your employer identification number (EIN). If you fail to provide this information in a timely manner, you may be subject to penalties.

You are not required to provide the information requested on a form that is subject to the Paperwork Reduction Act unless the form displays a valid OMB control number. Books or records relating to a form or its instructions must be retained as long as their contents may become material in the administration of any Internal Revenue law.

Generally, tax returns and return information are confidential, as required by section 6103. However, section 6103 allows or requires the Internal Revenue Service to disclose or give the information shown on your return to others as described in the Code. For example, we may disclose your tax information to the Department of Justice for civil and/or criminal litigation, and to cities, states, and the District of Columbia for use in administering their tax laws. We may also disclose this information to other countries under a tax treaty, to federal and state agencies to enforce federal nontax criminal laws, or to federal law enforcement and intelligence agencies to combat terrorism.

The time needed to complete and file these forms will vary depending on individual circumstances. The estimated average times are: **Form W-2**—30 minutes, and **Form W-3**—28 minutes. If you have comments concerning the accuracy of these time estimates or suggestions for making these forms simpler, we would be happy to hear from you. You can write to the Internal Revenue Service, Tax Products Coordinating Committee, SE:W:CAR:MP:T:T:SP, 1111 Constitution Ave. NW, IR-6406, Washington, DC 20224. **Do not** send Forms W-2 and W-3 to this address. Instead, see *Where to file* in the Instructions for Forms W-2 and W-3.

APPENDIX 16: DESCRIPTION OF CONTRACTOR CLASSIFICATIONS

Air and Water Balancing Contractor—An air and water balancing contractor installs any device and performs any work related to providing a specified flow of air in all types of existing heating and cooling systems and/or related to providing a specified flow of water in water piping systems.

Awnings Contractor—An awning contractor installs, modifies or repairs aluminum, metal, vinyl or canvas awnings and patio covers.

Boiler, Hot-Water Heating System Fitting Contractor—A boiler, hot water heating and steam fitting contractor installs, services and repairs power boiler installations, hot water heating systems and steam fitting, including fire tube and water tube steel power boilers and hot water heating low pressure boilers, steam fitting and piping, fittings, valves, gauges, pumps, radiators, convectors, fuel oil tanks, fuel oil lines, chimneys, flues, heat insulation and all other equipment, including solar heating equipment, associated with these systems.

Building Moving/Demolition Contractor—A building moving/demolition contractor raises, lowers, cribs, underpins, demolishes and moves or removes structures, including their foundations.

Cabinet, Millwork and Finish Carpentry Contractor—A cabinet, millwork and finish carpentry contractor makes cabinets, cases, sashes, doors, trims, nonbearing partitions and other items of "finish carpentry" by cutting, surfacing, joining, gluing and fabricating wood or other products to provide a functional surface. This contractor also places, erects, and finishes such cabinets and millwork in structures.

Central Vacuum Systems Contractor—A central vacuum systems contractor installs, modifies, maintains or repairs central vacuum systems, pneumatic tube dispatching systems or any other type of pipeline which operates systems of reduced pressure for any purpose.

Concrete Contractor—A concrete contractor forms, pours, places, finishes and installs specified mass, pavement, flat and other concrete work; and places and sets screeds for pavement flat work.

Concrete-Related Services Contractor—A concrete related services contractor installs reusable steel concrete form sections or interlocking precast pavers. Performs post-tensioning work, concrete sawing, breaking, curing, floor hardening treatment, coloring concrete, concrete restoration, coring work or operates a concrete pumping service.

Construction Cleanup Contractor—A construction cleanup contractor cleans up and/or removes from building grounds or structures any debris resultant from any construction project including but not limited to concrete, dirt, scrap lumber, plaster, drywall, any paint or adhesive products from windows, floors, ceramic tile and bathroom fixtures.

Construction Zone Traffic Control Contractor—A construction zone traffic control contractor prepares or removes lane closures, flagging, or traffic diversions, utilizing portable devices such as cones, delineators, barricades, sign stands, flashing beacons, flashing arrow trailers, and changeable message signs, on roadways, including but not limited to public streets, highways, or any public conveyance.

Doors, Gates and Activating Devices Contractor—A doors, gates and activating devices contractor installs, modifies or repairs all types of residential, commercial or industrial doors including overhead or sliding door assemblies.

Drilling, Blasting and Oil Field Work Contractor—A drilling contractor does core and post hole drilling, horizontal drilling and drilling for placement of charges and performing blasting work; performs drilling for site dewatering, oil well drilling and other oil field related specialty work.

Drywall Contractor—A drywall contractor lays out and installs gypsum wall board and gypsum wall board assemblies including nonstructural metal framing members, and performs the taping and texturing operations including the application of compounds that adhere to wall board to produce a continuous smooth or textured surface.

Earthwork and Paving Contractor—An earthwork and paving contractor digs, moves, and places material forming the surface of the earth, other than water, in such manner that a cut, fill, excavation, grade, trench, backfill or tunnel can be executed, including the use of explosives for these purposes.

Electrical Contractor—An electrical contractor places, installs, erects or connects any electrical wires, fixtures, appliances, apparatus, raceways conduits, solar photovoltaic cells or any part thereof, which generate, transmit, transform or utilize electrical energy in any form or for any purpose.

Electrical Sign Contractor—An electrical sign contractor fabricates, installs and erects electrical signs, including the wiring of such electrical signs.

Elevated Floors Contractor—An elevated floors contractor installs wood or metal framed elevated computer flooring systems.

Elevator Contractor—An elevator contractor fabricates, erects, installs and repairs elevators, including sheave beams, motors, sheaves, cable and wire rope, guides, cab, counterweights, doors, automatic and manual controls, signal systems, and all other devices and equipment associated with and efficient installation and operation of electrical, hydraulic and manually operated elevators.

Fencing Contractor—A fencing contractor constructs, erects, alters, or repairs all types of fences, corrals, runs, nailings, cribs, game court enclosures, guard rails and barriers, playground game equipment, backstops, posts, flagpoles, and gates, excluding masonry walls.

Flooring and Floor Covering Contractor—A flooring and floor covering contractor prepares any surface for the installation of flooring and floor coverings, and installs carpet, resilient sheet goods, resilient tile, wood floors and flooring, and any other materials established as flooring and floor covering material.

Fire Protection Contractor—A fire protection contractor lays out, fabricates and installs all types of fire protection systems; including all the equipment associated with these systems.

Framing and Rough Carpentry Contractor—A framing and rough carpentry contractor performs any form work, framing or rough carpentry necessary to construct framed structures; installs or repairs individual components of framing systems and performs any rough carpentry or associated work, including but not limited to the construction or installation of: sub-flooring, siding, exterior staircases and railings, overhead doors, roof decking, truss members, and sheathing.

General Building Contractor—A general building contractor is a contractor whose principal contracting business is in connection with any structure built, being built, or to be built, for the support, shelter and enclosure of persons, animals, chattels or moveable property of any kind requiring in its construction the use of at least two unrelated building trades or crafts, or to do or superintend the whole or any part thereof.

General Engineering Contractor—A general engineering contractor is a contractor whose principal contracting business is in connection with fixed works requiring specialized engineering knowledge and skill.

General Manufactured Housing Contractor—A general manufactured housing contractor installs, alters, repairs or prepares for moving any type of manufactured housing, including the accessory buildings or structures, and the foundations.

Glazing Contractor—A glazing contractor selects, cuts, assembles and/or installs all makes and kinds of glass, glass work, mirrored glass, and glass substitute materials for glazing; executes the fabrication and glazing of frames, panels, sashes and doors; and/or installs these items in any structure.

Hardware, Locks and Safes Contractor—A hardware, locks and safes contractor installs, modifies or repairs power and/or manually activated door and window locks with related hardware, built-in safes and vaults.

Hydroseed Spraying Contractor—A hydroseed spraying contractor applies seeds through any liquid media to any type of surface that has been prepared or contoured by others.

Insulation and Acoustical Contractor—An insulation and acoustical contractor installs any insulating media and preformed architectural materials for the purpose of temperature and/or sound control.

Landscaping Contractor—A landscape contractor constructs, maintains, repairs, installs or subcontracts the development of landscape systems and facilities for public and private gardens and other areas which are designed to aesthetically, architecturally, horticulturally, or functionally improve the grounds within or surrounding a structure or tract or plot of land. In connection therewith, a landscape contractor prepares and grades plots and areas of land for the installation of any architectural, horticultural and decorative treatment or arrangement.

Lathing and Plastering Contractor—A lathing and plastering contractor coats surfaces with a mixture of sand, gypsum plaster, quick lime or dehydrated lime and water, or sand and cement and water, or a combination of such other materials that create a permanent coating, including coatings for the purpose of soundproofing and fireproofing. A lathing and plastering contractor also applies and affixes wood and metal lath, or any other material prepared or manufactured to provide key or suction bases for the support of plaster coatings.

Lock and Security Equipment Contractor—A lock and security equipment contractor evaluates, sets-up, installs, maintains and repairs all doors and door assemblies, gates, locks and locking devices, panic and fire rated exit releases, jail and prison locking devices and permanently installed or built in safes and vaults.

Low Voltage Systems Contractor—A communication and low voltage systems contractor installs, services and maintains all types of communication and low voltage systems which are energy limited and do not exceed 91 volts. Such systems include, but are not limited to telephone systems, sound systems, cable television systems, closed circuit video systems, satellite dish antennas, instrumentation and temperature controls, and low voltage landscape lighting.

Machinery and Pumps Contractor—A machinery and pumps contractor installs, removes, modifies or repairs pumps; conveyors; cranes; dock levelers; various hoisting and material handling equipment; meters; dumb waiters; handicap lifting and assistance equipment; automated speedrail trolley systems for garment hanging; automatic car washing racks, and auto body paint finishing booth equipment.

Masonry Contractor—A masonry contractor installs concrete units and baked clay products; concrete, glass and clay block; natural and manufactured stone; terra cotta and fire brick or other material for refractory work.

Metal Products Contractor—A metal products contractor installs, modifies or repairs metal cabinets, lockers, modular storage structures, mail chutes, cable racks; aluminum or vinyl storm doors and windows; metal wall tiles; aluminum fascia covers; metal gutters; aluminum studs and trusses; metal railings and turnstiles; metal prison cell accessories such as welded-to-structure cell furniture; grills and cabinets.

Ornamental Metal Contractor—An ornamental metals contractor assembles, casts, cuts, shapes, stamps, forges, welds, fabricates and installs, sheet, rolled and cast brass, bronze, copper, cast iron, wrought iron, monel metal, stainless steel, and/or any other metal for the architectural treatment and ornamental decoration of structures.

Painting and Decorating Contractor—A painting and decorating contractor prepares by scraping, sandblasting or other means and applies any of the following; paints, papers, textures, fabrics, pigments, oils, turpentines, japans, driers, thinners, varnishes, shellacs, stains, fillers, waxes, adhesives, water and any other vehicles, mediums and materials which adhere by evaporation and may be mixed, used and applied to the surfaces of structures and the appurtenances thereto for the purposes of decorating, protecting, fireproofing and waterproofing.

Paperhanging Contractor—A paperhanging contractor applies all types and varieties of decorative wall coverings, except painting or paneling, including paper and vinyl goods, cork, burlap and carpet-type wall coverings.

Parking and Highway Improvement Contractor—A parking and highway improvement contractor applies and installs protective coatings, vehicle stops, guard rails and mechanical devices, directional lines, buttons, markers, signs and arrows on the horizontal surface of any game court, parking facility, airport, highway or roadway constructed of concrete, asphalt or similar material.

Pile Driving/Pressure Foundation Jacking Contractor—A pile driving and pressure foundation jacking contractor provides a pile driving and/or caisson drilling or auger service.

Pipeline Contractor—A pipeline contractor fabricates and installs pipelines for the conveyance of fluids, such as water, gas, or petroleum, or for the containment or protection of any other material, including the application of protective coatings or systems and the trenching, boring, shoring, backfilling, compacting, paving and surfacing necessary to complete the installation of such pipelines.

Plumbing Contractor—A plumbing contractor provides a means for a supply of safe water, ample in volume and of suitable temperature for the purpose intended and the proper disposal of fluid waste from the premises in all structures and fixed works.

Pole Installation and Maintenance Contractor—A pole installation and maintenance contractor installs wood or precast poles to support the wiring or cable that is installed by others or installs and maintains flag poles.

Pool and Spa Maintenance Contractor—A pool and spa maintenance contractor installs, replaces or repairs pool motors, pumps, filters, gas heaters and any above ground piping in connection with pools; includes electrical switches, breakers, pool lights, diving boards, existing solar systems that heat pools, pool and spa acid baths and applies vinyl liners to existing surfaces.

Prefabricated Equipment Contractor—A prefabricated products/equipment contractor performs installations of prefabricated products/equipment, including theater stage equipment; school classroom equipment; playground equipment; bleacher bench/seat component parts; store fixtures, and display cases; all forms and types of toilet/shower room partitions/accessories; prefabricated closet systems; laboratory and medical equipment, dust collecting systems; factory built fireplaces and accessories; major appliance installations and ventilating hoods in connection with existing fuel and energy lines which are installed by others; bus stop shelters; prefabricated phone booths; prefabricated sound proof environmental clean rooms; panelized refrigerated walk-in boxes; and all types of modular office, institutional or home improvement systems including, but not limited to all types of pre-finished and/or UL listed pre-wired wall panels.

Refrigeration Contractor—A refrigeration contractor constructs, fabricates, erects, installs, maintains, services and repairs refrigerators, refrigerated rooms, and insulated refrigerated spaces, temperature insulation, air-conditioning units, ducts, blowers, registers, humidity and thermostatic controls for the control of air, liquid and/or gas temperatures below fifty (50) degrees Fahrenheit, or ten (10) degrees Celsius.

Reinforcing Steel Contractor—A reinforcing steel contractor fabricates, places and ties steel mesh or steel reinforcing bars of any profile, perimeter, or cross-section, that are or may be used to reinforce concrete structures.

Roofing Contractor—A roofing contractor installs products and repairs surfaces that seal, waterproof and weatherproof structures.

Sand and Water Blasting Contractor—A sand and water blasting contractor uses the force of compressed air in conjunction with abrasive materials or water to clean or prepare surfaces for any protective, decorative and/or functional treatment.

Sanitation System Contractor—A sanitation system contractor fabricates and installs cesspools, septic tanks, storm drains and other sewage disposal and drain structures.

Scaffolding Contractor—A scaffolding contractor erects metal or wood scaffolding including temporary sidewalk sheltered construction work barricades.

Service Station Equipment and Maintenance Contractor—A service station maintenance contractor installs and/or removes underground fuel storage tanks up to 20,000 gallons which have been or are to be used for dispensing gasoline, diesel fuel, waste oil or kerosene.

Sheet Metal Contractor—A sheet metal contractor selects, cuts, shapes, fabricates, and installs sheet metal such as cornices, flashings, gutters, leaders, pans, kitchen equipment, duct work, including insulation, patented chimneys, metal flues, metal roofing systems and any other installations requiring sheet metal.

Siding and Decking Contractor—A siding and decking contractor applies or installs all types of exterior siding including wood, wood products, vinyl, aluminum and metal siding to new or existing buildings. This contractor also constructs wooden decks and related handrails.

Sign Installation Contractor—A sign installation contractor fabricates and installs all types of nonelectrical signs, including but not limited to: post or pole supported signs, signs attached to structures, painted wall signs and modifications to existing signs.

Solar Contractor—A solar contractor installs, modifies, maintains, and repairs active solar energy systems.

Structural Steel Contractor—A structural steel contractor fabricates and erects structural steel shapes and plates, of any profile, perimeter or cross-section, that are or may be used as structural members for buildings and structures, including the riveting, welding, rigging and metal roofing systems necessary to perform this work.

Suspended Ceilings Contractor—A suspended ceilings contractor installs, modifies or repairs all types of suspended ceilings, including but not limited to the following: lay-in-grid and other types of systems involving solid, perforated or translucent ceiling panels.

Swimming Pool Contractor—A swimming pool contractor constructs swimming pools, spas or hot tubs, including installation of solar heating equipment using those trades or skills necessary for such construction.

Synthetic Products Contractor—A synthetic products contractor installs synthetic counter tops and wall coverings; fiberglass, plastic, vinyl and epoxy products; plastic tile board and decorative art work; synthetic turf; bathtub and enamel refinishing, resin and epoxy application, and synthetic caulking and sealants; reservoir liners, vinyl swimming pool relining, pier piling wrap, and rodent guards; and PVC piping systems for irrigation and drainage; subsurface irrigation drip systems.

Tile Contractor (Ceramic and Mosaic)—A ceramic and mosaic tile contractor prepares surfaces as necessary and installs glazed wall, ceramic, mosaic, quarry, paver, faience, glass mosaic and stone tiles; thin tile that resembles full brick, natural or simulated stone slabs for bathtubs, showers and horizontal surfaces inside of buildings, or any tile units set in the traditional or innovative tile methods, excluding hollow or structural partition tile.

Tree Service Contractor—A tree service contractor prunes trees, removes trees, limbs or stumps, and engages in tree or limb guying.

Trenching Contractor—A trenching contractor is a limited specialty contractor whose contracting business is trenching only for foundations, pipelines, conduit and related trenching work.

Warm-Air Heating, Ventilation and Air-Conditioning Contractor—A warm air heating, ventilating and air-conditioning contractor fabricates, installs, maintains, services and repairs warm air heating systems and water heating heat pumps, complete with warm air appliances; ventilating systems complete with blowers and plenum chambers; air-conditioning systems complete with air-conditioning unit; and the ducts, registers, flues, humidity and thermostatic controls and air filters in connection with any of these systems.

Water Conditioning Contractor—A water conditioning contractor installs water conditioning equipment with the use of only such pipe and fittings as are necessary to connect the water conditioning equipment to the water supply system and to by-pass all those parts of the water supply system within the premises from which conditioned water is to be excluded.

Weatherization and Energy Conservation Contractor—A weatherization and energy conservation contractor installs, removes, modifies or repairs or provides maintenance services for energy conservation products limited to the following: door and window weather stripping, caulking, water heater pipe wrap, water heater blankets, insulating gaskets for electrical outlet covers, shade screens, shutters, storm windows, tinted window film, residential water flow restricting devices installed onto existing fixture.

Well Drilling Contractor—A well drilling contractor installs and repairs water wells and pumps by boring, drilling, excavating, casing, cementing and cleaning to provide a supply of uncontaminated water.

Welding Contractor—A welding contractor causes metal to become permanently attached, joined and fabricated by the use of gases and electrical energy, which creates temperatures of sufficient heat to perform this work.

Window Coverings Contractor—A window coverings contractor installs or applies decorative, architectural/functional window/glass treatment or covering products, including but not limited to all types of materials and fabrics that make up louvers, shutters, venetian and mini-blinds; residential or commercial draperies and screens; expanded metal window and door guards; plastic film window treatment and/or any other window treatment applied for temperature control or as a screening device.

Wood Tanks Contractor— Awood tanks contractor erects or repairs elevated wooden storage tanks and related cooling towers.

APPENDIX 17: HOMEOWNER/CONTRACTOR—CONTRACT FOR SERVICES

THIS AGREEMENT is hereby entered into this ___ day of__________, 20___ , between *[Name and Address]* ("Homeowner"), and *[Name and Address]* ("Contractor").

The above parties, for the consideration set forth below, hereby agree as follows:

1. Contractor agrees to build a deck for Homeowner in the exact dimensions and specifications set forth on the attached Exhibit A, which is made a part of this agreement.

2. Contractor agrees to provide and pay for all labor and materials necessary to complete the deck.

3. Homeowner agrees to pay Contractor the sum of *[Dollar Amount ($xxx)]* Dollars as and for compensation for labor and materials, payable as follows:

(a) *[Dollar Amount ($xxx)]* Dollars upon signing this agreement; and

(b) *[Dollar Amount ($xxx)]* Dollars when the deck is 50% completed; and

(c) *[Dollar Amount ($xxx)]* Dollars upon completion of the deck.

4. Contractor agrees that they will begin work on the deck within *seven (7)* calendar days from the date the contract is signed; and that the deck shall be completed within *forty-five (45)* calendar days from the date of contract.

5. Homeowner and Contractor agree that time is of the essence in completion of this deck, therefore, if the deck is not completed by the date set forth herein due to the neglect of the Contractor, Contractor agrees to pay Homeowner the sum of *[Dollar Amount ($xxx)]* Dollars

per day as liquidated damages until such time as the work is completed.

6. Notwithstanding Paragraph 5 of this agreement, in the event Contractor is delayed from completing the deck due to an act of God, fire, flood or other unavoidable event, or by fault of Homeowner, the date of completion shall be extended accordingly.

7. This contract contains the entire understanding between Homeowner and Contractor and any changes to this contract must be in writing and signed by both parties.

8. This contract is not assignable without the written consent of both parties.

9. This contract is governed by the laws of the State of *[Name of State]*.

10. The invalidity, in whole or in part, of any term of this agreement does not affect the validity of the remainder of the agreement.

____

[SIGNATURE LINE - HOMEOWNER]

____

[SIGNATURE LINE - CONTRACTOR]

APPENDIX 18: CONDITIONAL WAIVER AND RELEASE UPON PROGRESS PAYMENTS

CONDITIONAL WAIVER AND RELEASE UPON PROGRESS PAYMENT

Upon receipt by the undersigned of a check from [homeowner] in the sum of [dollar amount] payable to [subcontractor/supplier], and when the check has been properly endorsed and has been paid by the bank upon which it is drawn, this document shall become effective to release any mechanic's lien, stop notice, or bond right the undersigned has on the job of [homeowner] residing at [address], to the following extent.

This release covers a progress payment for labor, services, equipment, or material only and does not cover any retentions retained before or after the release date; extras furnished before the release date for which payment has not been received; extras or items furnished after the release date.

Rights based upon work performed or items furnished under a written change order which has been fully executed by the parties prior to the release date are covered by this release unless specifically reserved by the claimant in this release.

This release of any mechanic's lien, stop notice, or bond right shall not otherwise affect the contract rights, including rights between parties to the contract based upon a rescission, abandonment, or breach of the contract, or the right of the undersigned to recover compensation for furnished labor, services, equipment, or material covered by this release if that furnished labor, services, equipment, or material was not compensated by the progress payment.

Before any recipient of this document relies on it, said party should verify evidence of payment to the undersigned.

Date:

Authorized Signature: [Subcontractor/Supplier]

APPENDIX 19: UNCONDITIONAL WAIVER AND RELEASE UPON PROGRESS PAYMENTS

UNCONDITIONAL WAIVER AND RELEASE UPON PROGRESS PAYMENT

The undersigned has been paid and has received a progress payment in the sum of [dollar amount] for labor, services, equipment or material furnished to [homeowner] residing at [address].

The undersigned does hereby release any mechanic's lien, stop notice or bond right that the undersigned has on the above referenced job to the following extent:

This release covers a progress payment for labor, services, equipment, or material furnished only and does not cover any retentions retained before or after the release date; extras furnished before the release date for which payment has not been received; extras or items furnished after the release date.

Rights based upon work performed or items furnished under a written change order which has been fully executed by the parties prior to the release date are covered by this release unless specifically reserved by the claimant in this release.

This release of any mechanic's lien, stop notice, or bond right shall not otherwise affect the contract rights, including rights between parties to the contract based upon a rescission, abandonment, or breach of the contract, or the right of the undersigned to recover compensation for furnished labor, services, equipment, or material covered by this release if that furnished labor, services, equipment, or material was not compensated by the progress payment.

Date:

Authorized Signature: [Subcontractor/Supplier]

NOTICE: THIS DOCUMENT WAIVES RIGHTS UNCONDITIONALLY AND STATES THAT YOU HAVE BEEN PAID FOR GIVING UP THOSE RIGHTS.

THIS DOCUMENT IS ENFORCEABLE AGAINST YOU IF YOU SIGN IT, EVEN IF YOU HAVE NOT BEEN PAID. IF YOU HAVE NOT BEEN PAID, USE A CONDITIONAL RELEASE FORM.

APPENDIX 20: CONDITIONAL WAIVER AND RELEASE UPON FINAL PAYMENT

CONDITIONAL WAIVER AND RELEASE UPON FINAL PAYMENT

Upon receipt by the undersigned of a check from [homeowner] in the sum of [dollar amount] payable to [subcontractor/supplier, and when the check has been properly endorsed and has been paid by the bank upon which it is drawn, this document shall become effective to release any mechanic's lien, stop notice, or bond right the undersigned has on the job of [homeowner], residing at [address].

This release covers the final payment to the undersigned for all labor, services, equipment, or material furnished on the job [if applicable: except for disputed claims for additional work in the amount of (dollar amount)].

Before any recipient of this document relies on it, the party should verify evidence of payment to the undersigned.

Date:

Authorized Signature: [Subcontractor/Supplier]

APPENDIX 21: UNCONDITIONAL WAIVER AND RELEASE UPON FINAL PAYMENT

UNCONDITIONAL WAIVER AND RELEASE UPON FINAL PAYMENT

The undersigned has been paid in full for all labor, services, equipment or material furnished to [homeowner] on the job at [address] and does hereby waive and release any right to a mechanic's lien, stop notice, or any right against a labor and material bond on the job [if applicable: except for disputed claims for additional work in the amount of (dollar amount)].

Date:

Authorized Signature: [Subcontractor/Supplier]

NOTICE: THIS DOCUMENT WAIVES RIGHTS UNCONDITIONALLY AND STATES THAT YOU HAVE BEEN PAID FOR GIVING UP THOSE RIGHTS. THIS DOCUMENT IS ENFORCEABLE AGAINST YOU IF YOU SIGN IT, EVEN IF YOU HAVE NOT BEEN PAID. IF YOU HAVE NOT BEEN PAID, USE A CONDITIONAL RELEASE FORM.

APPENDIX 22: CONTRACT ARBITRATION/MEDIATION CLAUSES

CONTRACT ARBITRATION CLAUSE

Any controversy or claim arising out of or relating to this contract, or the breach thereof, shall be settled by arbitration administered by the American Arbitration Association in accordance with its Commercial Arbitration Rules, and judgment on the award rendered by the arbitrator(s) may be entered in any court having jurisdiction thereof.

CONTRACT MEDIATION CLAUSE

If a dispute arises out of or relates to this contract, or the breach thereof, and if said dispute cannot be settled through negotiation, the parties agree first to try in good faith to settle the dispute by mediation administered by the American Arbitration Association under its Commercial Mediation Rules, before resorting to arbitration, litigation, or some other dispute resolution procedure.

GLOSSARY

Alien—Any person not a citizen or national of the United States.

Base Rate Pay—An employee's basic hourly rate excluding overtime.

Blue-Collar Workers—Generally refers to individuals engaged in manual labor.

Bureau of Labor Statistics—A division of the U.S. Department of Labor that complies statistics related to employment.

Certificate of Citizenship—Identity document proving U.S. citizenship. Certificates of citizenship are issued to derivative citizens and to persons who acquired U.S. citizenship.

Certified Nurse's Aide (CNA)—Employee who assists elderly or incapacitated individuals with personal care and other activities of daily living.

Child Labor Laws—Network of laws on both federal and state levels, prescribing working conditions for children in terms of hours and nature of work which may be performed, all designed to protect the child.

Companionship—Employee who provides friendly visits and supervision to elderly or incapacitated individual. May also provide transportation and light housekeeping.

Consolidated Omnibus Budget Reconciliation Act (COBRA)—Federal law which extends a terminated employee's health insurance coverage at group rates for a specified period of time following termination.

Compensatory Time—Leave time given to an employee in lieu of premium pay for overtime work.

Cost of Living Allowance—An increase in wages which is based on an increase in the cost of living.

Deportation—The formal removal of an alien from the United States when the alien has been found removable for violating the immigration laws.

Double Time—Premium pay rate that is twice the employee's basic rate of pay.

Employability—The ability of an employee to meet the demands of a job and the conditions of employment.

Employer Sanctions—The employer sanctions provision of the Immigration Reform and Control Act of 1986 prohibits employers from hiring, recruiting, or referring for a fee aliens known to be unauthorized to work in the United States and subjects violators to a series of civil fines or criminal penalties.

Fair Labor Standards Act (FLSA)—Federal law governing federal wage and hour regulations.

Federal Unemployment Tax Act—Federal law which taxes employers for the purpose of funding the unemployment compensation program.

Home Health Aide—Employee who assists elderly or incapacitated individuals with bathing, dressing, feeding, and health care procedures that do not require professional nursing.

Housekeeper—Employee who assists with household chores, errands, transportation and meal preparation.

Immigration Reform and Control Act (IRCA) of 1986—Public Law 99-603 (Act of 11/6/86), which was passed in order to control and deter illegal immigration to the United States.

Immunity—As it relates to workers' compensation law, refers to the protection afforded an employer who is in compliance with workers' compensation statutes from common-law tort actions by employees arising out of the work-related injuries or illness.

Impairment—The loss, or loss of use of, any body part, system, or function.

Independent Contractor—An individual who contracts to perform services for others without qualifying legally as an employee.

Injury—Physical or mental harm or disability arising out of a work-related accident.

Labor Certification—Requirement for U.S. employers seeking to employ certain persons whose immigration to the United States is based on job skills, or nonimmigrant temporary workers coming to perform services for which qualified authorized workers are unavailable in the United States.

Labor Law—Legislation dealing with human beings in their capacity as workers or wage earners.

Layoff—A forced furlough from employment, on a temporary basis, generally caused by a lack of available work.

Loss of Earning Capacity—The impairment of ability to earn a living attributable to a work-related injury or illness.

Payroll Deduction—Money held from a worker's salary which is allocated to certain obligations, such as income taxes, union dues, health plan benefits, etc.

Payroll Period—The length of time between paydays.

Pension Plan—A retirement plan established by an employer for the payment of pension benefits to employees upon retirement.

Premium Pay—Wages paid to an employee above the basic pay rate, e.g. overtime pay.

Quid Pro Quo—Latin for "something for something." Refers to the exchange of promises or performances between two parties. Also refers to the legal consideration necessary to create a binding contract.

Scope of Employment—Those activities performed while carrying out the business of one's employer.

Self-Insured Employer—An employer that has been authorized by the appropriate agency to administer and pay directly on employee compensation claims.

Severance Pay—Monies paid to a terminated employee.

Shift—Refers to an employee's regular work hours.

Shift Differential—Premium pay earned by an employee who works an unusual or inconvenient shift, e.g. night shift.

State Insurance Fund—An insurance fund administered by the state instead of, or in addition to, private insurance.

Straight Time Pay—Wages paid to an employee for working his or her regular hours.

Take-Home Pay—Net wages paid to an employee after all applicable deductions are subtracted.

Temporary Employee—An employee who is hired to work on a short-term basis.

Temporary Worker—Under the Immigration Reform and Control Act of 1986, refers to an alien coming to the United States to work for a temporary period of time.

Termination—Refers to cessation of employment, e.g. by quitting or dismissal.

Vacation—Authorized leave from work, with our without pay.

Wages—Compensation paid to an employee.

Waiting Period—Statutory period of time, e.g. three days, from the date of work-related injury or illness, during which the employee is ineligible for workers' compensation payments, unless disability continues for a specified duration, e.g. two weeks.

Workers' Compensation—Refers to the benefits payable to claimants under a workers' compensation claim, such as lost wages, medical expenses, etc.

Work-related Injury or Illness—An injury or illness which is causally related to one's employment.

Wrongful Discharge—An unlawful dismissal of an employee.

Zone of Employment—The physical area in which injuries to an employee are covered by worker compensation laws.

BIBLIOGRAPHY AND ADDITIONAL RESOURCES

Black's Law Dictionary, Fifth Edition. St. Paul, MN: West Publishing Company, 1979.

Cornell Law School Legal Information Institute. (Date Visited: April 2005) <http://www.law.cornell.edu/>.

Insurance Information Institute. (Date Visited: April 2005) <http://www.iii.org/>.

Internal Revenue Service (Date Visited: April 2005) <http://www.irs.gov/>.

The Society for Human Resources Management (Date Visited: April 2005) <http://www.shrm.org/>.

U.S. Citizenship and Immigration Services (Date Visited: April 2005) <http://uscis.gov/>.

The U.S. Equal Opportunity Commission (Date Visited: April 2005) <http://www.eeoc.gov/>.

U.S. Department of Labor (Date Visited: April 2005) <http://www.dol.gov/>.

DATE DUE

GAYLORD			PRINTED IN U.S.A.